ELECTRONIC PUZZLES

AND GAMES

MATTHEW MANDL

Lecturer in Electronic Technology
Technical Institute & Community College
Temple University
Philadelphia, Pa.

TAB BOOKS
BLUE RIDGE SUMMIT, PA. 17214

First Printing — May, 1958
Second Printing — October, 1959
Third Printing — January, 1961
Fourth Printing — May, 1962
Fifth Printing — February, 1965
Sixth Printing — March, 1966
Seventh Printing — June, 1968

Library of Congress Catalog Card No. 58-11148

contents

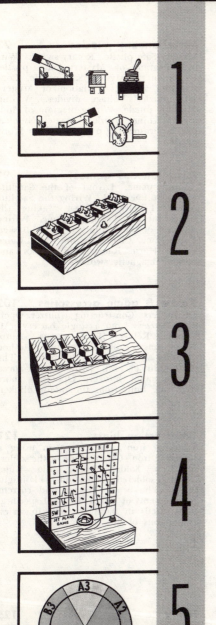

Switches 7
Connecting switches. Wire size. Single-pole single-throw knife switch. Single-pole double-throw knife switch. Pushbutton switch. Toggle switch. Wafer switch. The metal-plunger switch. How to make a metal-plunger switch. How the metal-plunger switch works. The wood-plunger switch. How the wood-plunger switch is used.

Simple puzzles 17
Knife-switch puzzle. How to construct the knife-switch puzzle. Wiring diagram of the knife-switch puzzle. Making the puzzle more complicated. Double-indicator puzzle. Wiring the double-indicator puzzle. Double-indicator puzzle construction. Game puzzle. Wiring diagram of the game puzzle. How to win the game puzzle.

Advanced puzzles 27
The street-light puzzle—an advanced type of switch puzzle. Wiring diagram of the Street-Light puzzle. Construction of the Street-Light puzzle. River-Crossing puzzle. Switch construction details for the River-Crossing puzzle. The Little Thinker. Twenty-one.

Games for two players 47
Hide-and-Seek game. Physical layout of the Hide-and-Seek game. Trip-to-the-Shore. Top view of the Trip-to-the-Shore. Wiring diagram of the Trip-to-the-Shore. Making a spinner dial. Marking and mounting the spinner dial. Playing instructions Trip-to-the-Shore game. Jet Plane game. Building the Jet Plane game.

Games for several players 63
Cross-Country Race. Variable and fixed indications. Top view of the Cross-Country Race. Wiring diagram of the Cross-Country Race. High- and low-speed spinners. Space-Travel game. Wiring diagram for the Space-Travel game. Construction of the Space-Travel game. Space-Travel spinner. Playing instructions.

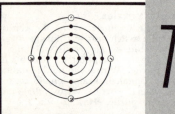

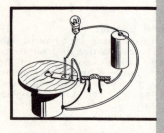

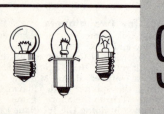

Miscellaneous puzzles 73

Mystery Safe puzzle. Rotary switch. Wiring diagram of Mystery Safe puzzle. Layout of the Mystery Safe puzzle. Constructing a dial. Variation of Mystery Safe puzzle. Voltage divider. Adding Machine puzzle. Wiring diagram of Adding Machine puzzle. Magic Number puzzle. Prospecting puzzle. Hidden Word puzzle.

Miscellaneous games 91

Satellite game. Layout of the Satellite game. Spinner dial. Wiring the Satellite game. Tri-target game. Making the switch for the Tri-target game. Wiring diagram of the Tri-target game. Basket-Catch game. Chassis for the Basket-Catch game. Horse Race game. Series and parallel magnetic switches.

Puzzle & game accessories 105

DC buzzers. Constructing a buzzer. Operating buzzers from a single battery. AC buzzers. Relays. Delay systems. Automatic camera cable-release units. Method of using an automatic cable-release. The motor blinker unit. Reducing the speed of the motor blinker. Operating more than one bulb from the motor blinker.

Construction hints 121

Soldering iron. Precautions in using a soldering iron. Care and use of the soldering iron. Solder. Solder ratio. Rules for well-soldered connections. Flashlight bulbs. The bead. Voltage and current requirements of bulbs. Dry cell specifications. Cells in parallel. Cells in series. Dimensions of dry cells.

Index 125

introduction

THE person who likes to build electrical devices as a hobby will find the construction of puzzles and games employing electrical principles particularly fascinating. The puzzles and games contained in this book are both simple and complex. Those who have had limited experience in working with circuits will find the puzzles and games in the earlier chapters particularly interesting because they are not only easy to build but will serve as an introduction to the methods employed for constructing the more advanced types. The simple puzzles require perhaps only a half hour of construction time while the more complex become a worth-while project involving several evenings of the hobbyist's time. To facilitate their construction by the newcomer interested in electronics as a hobby, layout drawings have been included for each puzzle or game as well as a photograph of the basic type of puzzle discussed.

The person experienced in the assembly of circuits may make a random selection of any particular puzzle or game since each discussion is complete in itself. For the most part the games and puzzles chosen for inclusion in this book are of such a nature that no special electrical devices are necessary. For the same reason, equipment needed for construction of the puzzles is limited to such items as tin shears, soldering iron and a few basic woodworking tools for fabricating the wooden chassis which are used.

Once the reader starts on the construction of one of these puzzles or games he will not only derive the pleasure of building an electrically operated device, but will also find that the completed product will provide hours of amusement for grownups as well as for youngsters. Eventually, if all the games and puzzles in this book are assembled, they will make a welcome addition to the gameroom, not only for youngsters on a rainy day, but for the amusement of everyone at gatherings and parties.

Yardley, Pa.   MATTHEW MANDL

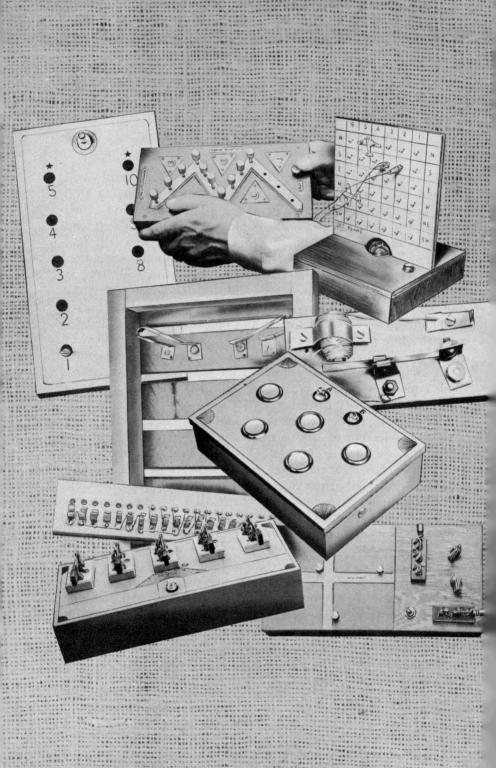

switches

S WITCHES are probably the most important item in puzzles and games which employ electrical or electronic principles in their construction. Switches are used either to close contacts to permit an electric current to flow into a certain section of the puzzle or game, or to open the circuit to prevent the current from flowing to a particular part. Because switches are the heart of any electronic puzzle or game, some of the basic types are described in this chapter. Thus, in the construction of the puzzles and games detailed later, reference can be made to this chapter for a description of the particular switch or switches involved. In some instances a choice can be made between one or two switch types and this will facilitate construction of the puzzle or game because the builder can select the more convenient switch type.

Besides switches, most of the games and puzzles in this book employ only flashlight bulbs and batteries[1] so that they can be constructed without having to acquire special parts. Once the unit has been built, additional refinements can be added as detailed and illustrated in the chapter on accessories.

Connections between switches are made with ordinary hookup wire, although virtually any type of wire can be employed. Best for interconnecting switches is the push-back type where the insulation can be moved back from the ends. This eliminates

---

[1] Single batteries may be used in these puzzles and games without regard to polarity.

the necessity of stripping off the insulation from the ends prior to attaching or soldering the wire ends to the switch contacts.

Because very little current flows in the puzzle circuits, hookup wire can be as small as No. 24 or No. 26. On the other hand, however, larger-size wire such as No. 18 bell wire could also be employed if it is handy. Bare wire could also be used; the only precaution to observe is that the various wires of any unit do not touch each other. Thus, if it is necessary to cross one wire over another, one of the wires should be wrapped with electrician's

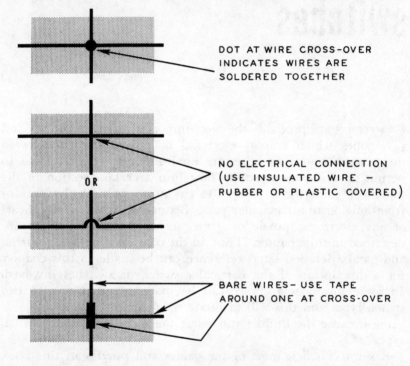

DOT AT WIRE CROSS-OVER INDICATES WIRES ARE SOLDERED TOGETHER

NO ELECTRICAL CONNECTION (USE INSULATED WIRE — RUBBER OR PLASTIC COVERED)

BARE WIRES— USE TAPE AROUND ONE AT CROSS-OVER

OR

Fig. 101. *Wires to be joined are shown as a heavy dot. If the dot isn't shown, then the wires are not to be joined. Bare wires should be covered with some form of insulating material to prevent accidental contact.*

tape to prevent electrical contact and a consequent short circuit (see Fig. 101).

When it is necessary to solder wires to switches, it is preferable to use tinned wire to expedite the soldering process. If the wire is untinned, the end of the wire which is to be soldered to the switch must be scraped with a knife and a good grade of rosin-core solder applied.

Other than these few precautions, construction of the puzzles

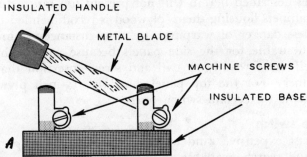

INSULATED HANDLE

METAL BLADE

MACHINE SCREWS

INSULATED BASE

A

Fig. 102-A. *The illustration at the top shows a single-pole single-throw knife switch. Connections are made to machine screws. Mounting holes for wood screws are in the base of the switch.*

Fig. 102-B. *The drawing at the right is that of a single-pole double-throw switch. This switch has three connections—one at each end and also to the blade. (Blade connection not shown in this drawing.)*

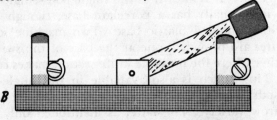

B

Fig. 102-C. *The pushbutton switch, shown at the left, closes the circuit only so long as the button is pressed. The switch opens when there is no pressure on the pushbutton. This switch has only two connections.*

PUSHBUTTON

FLANGE

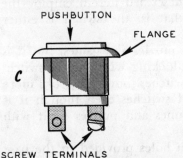

C

SCREW TERMINALS

Fig. 102-D. *A toggle switch is shown at the right. The action of this switch is very much like that of the single-pole single-throw switch. The clearance hole for mounting the switch is between 3/8 to 1/2 inch.*

MOUNTING NUT

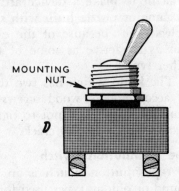

D

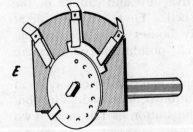

E

Fig. 102-E. *The drawing at the left shows a simple type of wafer switch. The moving portion is known as the rotor. Connections are soldered to the fixed contacts. Rotary switches are more complicated than the other types on this page, but they can also perform many more switching functions.*

9

and games described herein will not present too much difficulty. For the cabinets housing them, plywood is recommended because there is less danger of warping. In most instances ½-inch plywood is desirable for the side panels because such a thickness lends itself more readily to nail and screw insertion than does ¼-inch stock. For the top panel, however, ¼-inch plywood is more suitable for certain switch types.

The knife switch

The least expensive kind of switch which can be purchased is the knife switch available in 5-and-10¢ stores and hardware stores. The most simple knife switch is the one shown at A of Fig. 102. It is known as the single-pole–single-throw (spst). This switch usually has a porcelain base, though some are available with a molded plastic base. Two machine screws for attaching wires are provided, one at the base of the swivel section and the other where the movable knife section makes contact when closed.

This switch is also available in the single-pole–double-throw (spdt) type—as illustrated at B—as well as the double-pole–double-throw variety. The latter is mentioned only for reference purposes, since almost all the puzzles described herein employ the single-pole–single-throw principle; that is, the switch is either open or closed.

The knife switch can be used for puzzles and games, but it is difficult to make a device attractive-looking with such a switch. A game or puzzle built with such switches, however, looks more "electronic" because of the exposed switches even though it is not as attractive as some of the games and puzzles built with other types.

The knife switch has two through holes provided in the base to accommodate wood screws or bolts for mounting purposes. Care must be taken not to tighten the screws or bolts too much because the base may crack.

The pushbutton switch

A pushbutton switch is an attractive unit and can be of the round or oblong types used for doorbells. These are available at hardware stores or electrical supply houses. While more impressive than the simple knife switch, pushbutton switches are usually more costly.

The pushbutton switch is a simple make-and-break type in which a contact is made when the button is depressed. The contact is opened when pressure on the button is released. Two screw terminals are provided at the bottom of the switch for facilitating the connection of the wires. The pushbutton switch is essentially a single-pole–single-throw type, though special-purpose and more elaborate kinds are available. The basic pushbutton switch is illustrated at C of Fig. 102. This type of switch is mounted by drilling in the panel a hole having a diameter slightly larger than the switch. Some of these pushbutton switches have a small flexible flat spring at the sides which holds the switch tight after it is pushed into place. The flange at the top of the switch rests on the panel top to make a flat installation.

The toggle switch

A variety of toggle switches is available, some consisting of the smaller types used in the older radios, electric train sets and other such devices. Other types are those found in wall switches of home lighting systems. The simple toggle switch shown at D of Fig. 102 is of the single-pole–single-throw type. When the toggle is pushed in one direction, the switch contacts close; when the toggle is pushed in the other direction, the switch contacts open.

The top of the switch has a threaded section with a nut for mounting the switch to a panel. A hole must be drilled in the panel, making sure the hole has sufficient clearance for insertion of the threaded sections of the toggle switch. The large-hole nut is then turned into place for holding the switch. In most of the smaller toggle switches, the wires must be soldered to the terminals provided. For larger toggle switches, such as wall switches in homes, screw terminals are available.

Rotary switches

These switches consist of two parts; a fixed section known as the stator and a revolving section called the rotor. Rotary switches can be fairly complex and in addition may not have the durability of the switches described earlier. However, they can be used for puzzles and games. A simple type of rotary switch is shown in E of Fig. 102.

The metal-plunger switch

A quite simple but effective switch especially suitable for games and puzzles can be constructed from a metal rod and strips of tin. Such a switch is not only easy to construct but is virtually foolproof as well as quite long-lasting. This type of switch is illustrated in Fig. 103. The metal rod is a 2-inch section cut from either a curtain rod, an aluminum antenna element or other similar tubular metal. The switch becomes part of the panel section of the puzzle or game and for best results the top panel used for this type of switch should be at least ½ inch thick.

A hole having a diameter slightly larger than the plunger is drilled into the panel. The hole should not be so large that the plunger is loose when it is inserted. It is better to have the hole of such a size that the plunger may be inserted easily without binding, but yet be held in a fairly rigid upright position. One

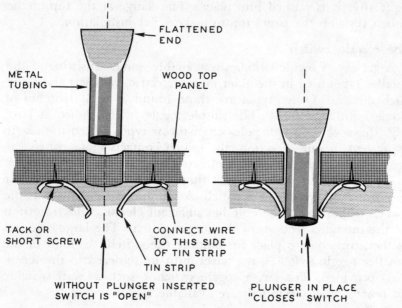

Fig. 103.   *The plunger switch is easy to make and is practically fool-proof. The plunger can be made of any type of metal tubing. The tubing should be clean and should have no paint on it. For good contact, clean the tubing with steel wool. The clearance hole in the base should be slightly larger than the plunger.*

end of the 2-inch plunger should be flattened for about ½ inch by placing it in a vise. Heavy pliers may also be used for flattening the end, or it can be hammered flat. The flattened end acts as a stop when the plunger is inserted into the hole and prevents the plunger from falling through. The flattened portion can be used for mounting identifying tags on each plunger or for other purposes as dictated by the particular game or puzzle used.

The top of the hollow metal plunger can also be used for screw-mounting of disks, circular or square wood sections, or other material such as plastic. The metal plunger is prepared for such mounts by hammering wood into the top portion, as shown in

Fig. 104. Stand the metal plunger on a vise or other hard surface, and hammer a piece of wood (slightly larger in diameter than the plunger) over the top hole, with the grain of the wood running in the same direction as the length of the metal plunger. The result will be a tight insertion of a wooden slug into the top of the metal plunger. This provides a means for mounting a section to the top by use of a wood screw as shown. A wooden dowel can also be used. Make a force fit into the metal tube.

As shown in Fig. 103, the plunger makes contact with two metal strips underneath the panel. These contact strips are cut from

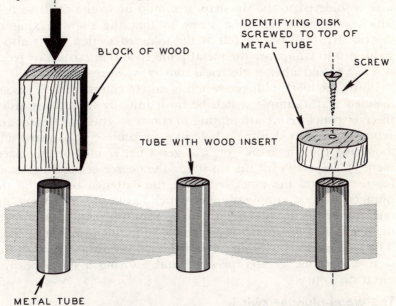

IDENTIFYING DISK
SCREWED TO TOP OF
METAL TUBE

SCREW

BLOCK OF WOOD

TUBE WITH WOOD INSERT

METAL TUBE

Fig. 104. *Plunger switches can be made from scrap pieces of wood and metal tubing. Use soft wood. After the wood has been inserted in the tubing, you can then screw a disk to the top of the plunger. The disk can be sanded and painted some attractive color.*

tin cans or from thin sheet tin such as used for roofing. The author has found that the tin from tin cans is particularly suitable because, being already tinned, it is easy to solder wire to the strips. Also, short sections of such tin strips are sufficiently flexible to make good contact without applying too much pressure to the plunger.

The strips can be cut approximately 1¼ inches long by ¼ inch wide. A hole is then punched near one end of this tin strip using an icepick or an awl. Lay the tin strip over a piece of soft wood and stab the icepick into the tin. A carpet tack or a ½-inch wood screw can then be used to fasten the tin to the underside of the panel. The tin strip is bent before mounting, so that the bent section will make contact with the plunger when the latter is inserted. Each plunger switch requires two tin strips, one on each side of the plunger as shown in Fig. 103. The tin strips should be so bent that there is a slight gap between them. Thus, the switch is "open" with the plunger out but, when the plunger is inserted, it closes the switch by making contact with both tin strips. If the tin strips are almost touching, the insertion of the plunger forces the tin strips out slightly and thus a good contact is insured. When wire is soldered to the tin strip, it should be to the small section which extends beyond the screw so that the slight flexing of the tin strip will not be felt at the soldered section. It is also a good idea to sandpaper the metal plunger so that tarnish or paint is removed and a better electrical contact is assured.

While the metal-plunger switch is easy to construct, it is recommended that a sample switch be built initially from a discarded piece of panel before attempting to construct the switches for the actual puzzle. By drilling a hole in a scrap piece of paneling and mounting the tin strips, you will get a better idea of the placement of the screws for the tin strips, the degree of bend necessary for the tin and the exact length of the flattened portion of the plunger. Once a sample plunger switch has been constructed, the ones for the puzzle will be found to be a routine matter. A plunger diameter of 1/4 to 1/2 inch is preferable to tubing less than 1/4 inch. The larger-diameter tubing bends the strips outward to a greater degree, exerting greater pressure and assuring a more positive electrical contact.

The wood-plunger switch

A switch similar to the metal-plunger type can be constructed from wood plungers. The tin strips beneath the chassis are bent so that they touch each other, thus making electrical contact *when the plunger is removed.* When the wood plunger is inserted, it pushes the two tin strips apart and thus opens the circuit. Hence this type of switch works in opposite fashion to the metal-plunger switch.

For the wood-plunger switch, wooden dowels having a diameter of 1/4 inch or slightly smaller are preferred. Larger dowels will push the tin strips too far apart and may reduce the tension present at the tin contact points when the plunger is removed. The manner in which the tin can be bent for maximum flexibility is shown in Fig. 105. Dowels of various sizes are available at lumber yards and are also sold in bundles by some retail mail-order houses. Another source is in toy departments of stores, where boxes of various-sized "sticks" can be purchased.

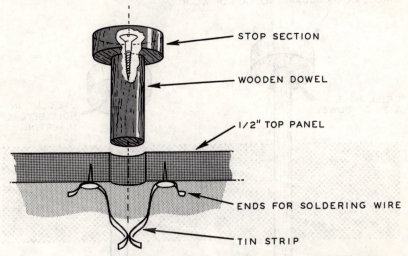

STOP SECTION

WOODEN DOWEL

1/2" TOP PANEL

ENDS FOR SOLDERING WIRE

TIN STRIP

Fig. 105. *This illustration shows a wood-plunger switch. This type of switch works in a manner exactly opposite to the metal-plunger switch described earlier. Any convenient size dowel can be used. The stop section or disk is the same as that used on the metal-plunger switch.*

As with the metal plungers, some sort of stop or top mount can be attached to the plunger section which remains above the panel. Appropriate identifying symbols or illustrations can then be glued to the top section as required by the particular puzzle.

With dowels of small diameter, the top mount or stop cannot be added very easily with a screw such as shown for the metal plunger switch. The small diameter of the wooden dowel would not provide sufficient body for the insertion of a screw from the top without splitting the dowel. Also, because of the small diameter, any attempt to glue a flat section to the top of the dowel would fail because the slightest pressure on the top would disengage the glued joint. The best method for assuring a tight top section which will act as a stop to prevent the dowel from falling into the plunger hole is that shown in Fig. 106. A square or circular section of wood is cut out to act as the top mount. This section of wood can be approximately ½ inch in diameter (or ½ inch square) and about ½ inch thick. A hole is then drilled at its center, about half way through so that the dowel can be inserted. The dowel will fit snugly into the hole to insure a good connection. Do not make the hole too small, however, or the wood will split as the dowel is forced into it. Apply some glue to the end of the dowel before inserting it. Press it in place into the hole. When the glue has dried, a firm top mount

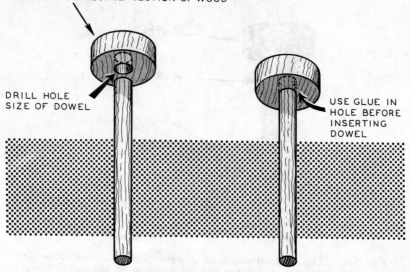

Fig. 106. *Method for mounting a stop section on a thin wooden dowel. The dowel should fit snugly into the hole.*

has been constructed. The top of this mount can then be identified by coloring it or by numbering it in accordance with the instructions given later for the particular puzzles or games involved.

Switch materials

The tin strips for the plunger switches are readily obtained from discarded tin cans. The phrase tin cans is not quite correct since the cans are made of steel with a thin coating of tin. This makes the metal ideal for use as a magnetic switch, as shown on page 83.

Not all tin-coated steel [used in tin cans] is suitable. Before building a complete game, experiment by making a plunger switch and observe the springiness of the metal. It may be necessary to try several different cans before suitable spring metal can be obtained.

Phosphor bronze and spring brass can also be used, but these metals are not too readily available. With a little ingenuity, however, spring metal can be obtained from unexpected sources. It is found in some types of weather stripping. Abandoned toys and games often contain spring metal which can be salvaged. Sometimes spring metal can be purchased from hobby shops, locksmiths, hardware stores or lumber yards.

simple puzzles

SIMPLE puzzles can easily be constructed using a minimum amount of materials. It is always advisable for the amateur constructor to start with such puzzles to gain experience and confidence. However, even the easiest puzzle described here can readily be made more complex through the addition of extra components.

Knife-switch puzzle

One of the easiest puzzles to construct is the one shown in Fig. 201. Here, five knife switches are mounted on a wooden chassis which also has a flashlight bulb mounted at the center front for indicating when the puzzle is solved. In this puzzle, all switches are initially in the open position. The problem is to cause the flashlight bulb to light by closing certain switches. The person solving the puzzle does not know whether a single switch may light the bulb or whether a combination of two or more switches must be closed. Because of the variety of combinations, the chances of selecting the proper switches initially are remote.

The puzzle shown in Fig. 201 was built on a chassis measuring 9 inches wide, 4 inches deep and 1/2 inch high. For the top panel 1/4-inch plywood was used and 1/2-inch stock for the sides. The switches are mounted in a row as shown. If the switch bases also have holes for the wiring, such holes can be marked on the panel with a pencil before the switch is mounted, and holes drilled into the top panel so that the wires can be run directly from the screw terminals through the base of the switch and to the underside of the chassis. If the switch bases have holes only for the mounting screws, and not for the wires, small holes will have to be drilled alongside each switch for feeding through the wires to the underside.

A hole is also drilled at the front center of the panel to accommodate the flashlight indicator bulb. The hole for the flashlight bulb should be of such a size that the bulb fits rather snugly and is thus held rigidly. No flashlight socket is needed because the wires are soldered directly to the flashlight bulb. If desired, a

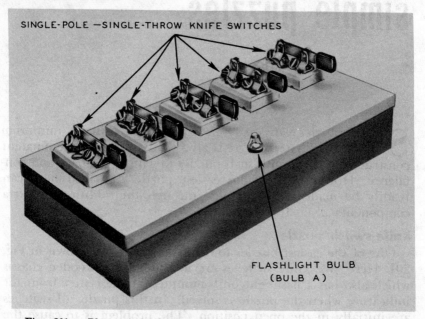

Fig. 201.   *Photo of the knife-switch puzzle. Each switch requires two holes for the connecting wires to go through. Mark and drill these holes before mounting the switches. The switches are held in place with wood screws. You can connect wires to the terminals on these switches without soldering.*

small type of socket can be used and can easily be purchased at a hardware store or at a radio wholesale house. Because the bulbs will last a long time, however, the wires can be soldered directly to the latter, one wire being connected to the center contact point of the bulb and the other to the threaded brass side. The wiring diagram for this puzzle is shown in Fig. 202.

In this particular puzzle, the first, fourth and fifth switches are wired in series as shown in the diagram. Series wiring means that a wire goes from one switch to the other in succession as shown. Switches 2 and 3 are wired in parallel; that is, they are wired across the flashlight bulb marked A and when either of these switches is closed it will short out bulb A and prevent it from lighting. Thus, the second bulb B is for protection purposes should all the switches be closed at once. If this happens, flashlight bulb B prevents the battery from being short-circuited.

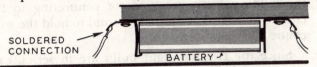

SOLDERED
CONNECTION

BATTERY

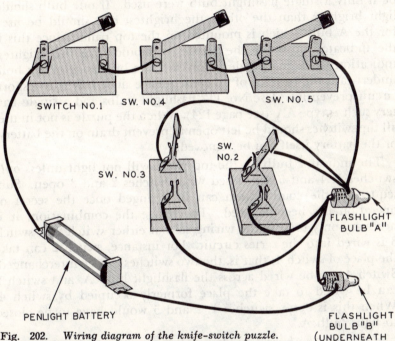

SWITCH NO.1   SW. NO.4   SW. NO. 5

SW.
NO. 2

SW. NO.3

FLASHLIGHT
BULB "A"

PENLIGHT BATTERY

FLASHLIGHT
BULB "B"
(UNDERNEATH
CHASSIS)

Fig. 202. *Wiring diagram of the knife-switch puzzle. Two flashlight bulbs are used. The illustration at the top shows how the battery is inserted in its holder. The battery can be mounted in any position.*

The wiring can consist of almost any type of wire available around the house since very little current is carried by the wires. If any of the wires cross each other, insulated wire should be used. Bare wire can be employed if care is taken not to have the bare wires touch each other.

Clips for the battery[2] can be cut from a tin can or a sheet of

---

[2] In this book the words cell and battery are used interchangeably. Actually, a cell is a single unit. A battery consists of one or more cells.

roofing tin. Cut a strip approximately ½ inch wide and 2½ inches long. Bend this around so that it will hold the cell as shown in Fig. 202. Punch a hole near the end away from the battery and fasten this tin strip to the underside of the top panel with a small wood screw. The ends of the tin extending beyond the holding screw are used for connecting to the wires of the circuit. The tin can be bent around to hold the wires or a soldered connection can be made.

Since the two flashlight bulbs are in series, each should light with equal brilliancy though neither will be as bright as it would be if only a single flashlight bulb were used. If one bulb should light brighter than the other, the brightest one should be used for the A bulb which is mounted on the top panel; since this is the indicator bulb for the puzzle, it should give the brightest indication when the puzzle is solved. The brilliancy of the bulb underneath the panel is of no consequence since it is only a short-circuit preventive. Use No. 112 flashlight bulbs and a single battery such as type AA (see page 124). When the puzzle is not in use, all the switches should be left open to prevent drain on the battery, or the battery itself can be removed.

The indicator bulb on the top panel will not light unless *only* switches 1, 4 and 5 are closed with switches 2 and 3 open. Subsequently, the combination can be changed once the secret of operation has been learned. To change the combination, it is necessary only to alter the wiring so that either switch 2 or switch 3 is wired into the series circuit. For instance, switch 2 can take the place of switch 4; that is, the two switches can be interchanged. Switch 4 can be wired across the flashlight bulb A, and switch 2 can be wired to take the place formerly occupied by switch 4. When this is done, switches 1, 2 and 5 would have to be closed to light bulb A.

If desired, the puzzle can be made more complicated by adding one or two additional switches to those already in use. Such extra switches can be wired with dummy wires attached to them but not connected to the puzzle circuit. Since the person solving the puzzle does not know which switches are actually in the circuit and which are not, the trial-and-error procedures are lengthened considerably and the puzzle becomes more complex.

Instead of knife switches, pushbutton switches of the type described in the first chapter can also be used. The pushbuttons make a neater-appearing puzzle but are more costly than knife switches. When using pushbutton switches, follow the wiring shown in Fig. 202 for the knife switch. If desired, the puzzle can also be constructed using toggle switches or either of the home-made switches shown in Fig. 103 or Fig. 105.

Double-indicator puzzle

Another version of an easily built switching type puzzle is that shown in Fig. 203. In this puzzle, two indicator bulbs are visible on the top of the panel as shown. This particular unit was built with pushbuttons to illustrate their appearance in puzzles of this type though knife switches or any of the other kind described in Chapter 1 can be employed instead.

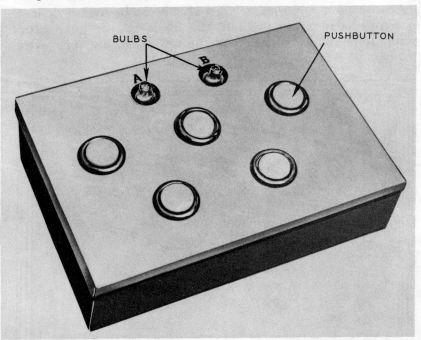

BULBS   PUSHBUTTON

Fig. 203. *This type of puzzle is known as a double indicator. It uses five pushbutton switches and two light bulbs. All of the push-buttons are active (connected). The puzzle can be made more com-plicated by adding pushbuttons which are not connected into the circuit. Other types of switches can be used if so desired.*

For the puzzle shown in Fig. 203, the problem consists of find-ing the combinations of pushbuttons which, when depressed, will light bulb B alone, and also finding the combinations of buttons which will light both bulbs simultaneously. Fig. 204 shows the wiring diagram for this two-indicator puzzle: pushbuttons 1, 3 and 5 must be depressed to light both bulbs. To light bulb B alone, pushbuttons 1, 3, 4 and 5 or pushbuttons 1, 2, 3 and 5 must be depressed. As with the previous puzzle, 1/4-inch plywood can be used to construct the small chassis of the puzzle. For the unit illustrated in Fig. 204, the chassis was built to measure 5 1/2 inches long, 4 inches wide and 1 1/2 inches high. Tin strips are used to form holders for the battery. For the flashlight bulbs holes are drilled just large enough so that the bulbs will press into place and remain there because of the tight fit. The flashlight bulbs are pressed into place *after* thin wires are soldered to the side and base of the bulbs. (No. 112 bulbs are recommended. See page 124.)

As with the first puzzle, a larger chassis can be used if desired to accommodate more pushbuttons. Thus, two or three extra pushbuttons can be included in the top panel, but such extra

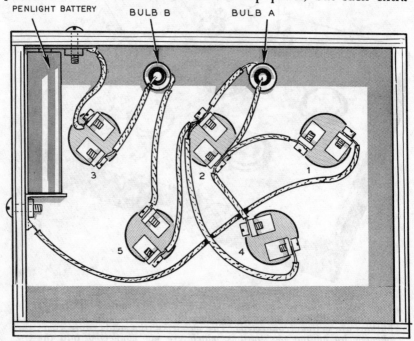

Fig. 204. *Wiring diagram of the double-indicator puzzle. You can solder directly to the light bulbs or you can use sockets for them. Very little soldering is required since most of the connections are to machine screws mounted on the pushbutton switches. A single penlight battery is used.*

pushbuttons are not connected into the main puzzle circuit. They act only as dummies. Their addition to the five pushbuttons shown in Fig. 203, however, will lengthen the time it takes to find the proper combination to light both flashlight bulbs.

Game puzzle

A puzzle employing plunger type switches is shown in Fig. 205.

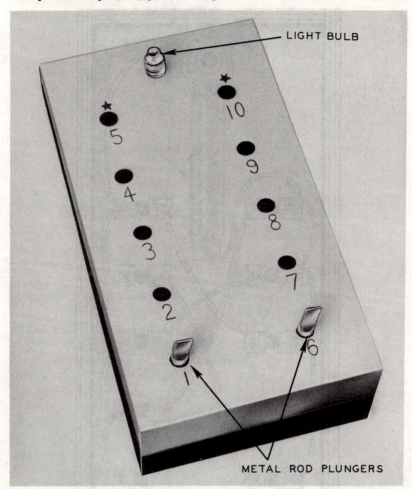

LIGHT BULB

METAL ROD PLUNGERS

Fig. 205. *This game uses metal-rod plungers of the type described in the first chapter. Only two plungers are required to play this game. The plungers in the photograph shown above are in their starting position. Dummy holes can be drilled to make the game more complicated.*

As shown, the top panel has 10 holes, numbered from 1 to 10. Two metal-rod plungers are used in this puzzle, these plungers being inserted at the bottom of the board, in holes 1 and 6 as shown. This puzzle is also a basic game since the user is actually playing against the device in trying to solve the puzzle. The object of the game is to move either of the plungers progressively up the board until the left plunger has reached hole number 5 and the right plunger has reached hole number 10. Only one

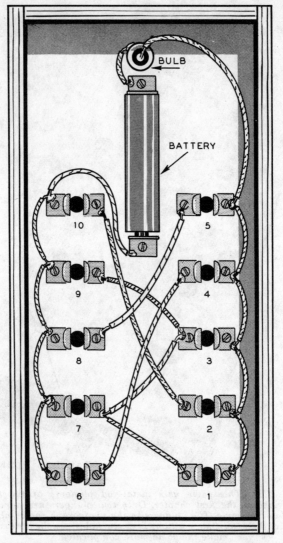

Fig. 206. *This is the wiring diagram of the puzzle. Since many wires cross each other it is advisable to use insulated wire. Scrape and tin the ends of the wires before soldering. Make sure the plungers make firm contact with their tin strips on the underside of the base.*

plunger may be moved at a time and, if the indicator bulb lights, both plungers should be returned to their starting point.

In this game, cross wiring between the holes is utilized so that the results of placing a plunger in a particular hole will vary from time to time. Thus, a plunger in holes 6 and 1 will not light the bulb, yet a plunger in hole 6 will light the bulb, provided the other plunger is in hole 4. There are only certain moves which can be made to prevent the bulb from lighting by the progress of the plunger up the board. For children in the younger age groups, the puzzle can be simplified by permitting the game to progress, even when the bulb lights, by simply returning the plunger which caused the bulb to light back to the position preceding the wrong move and then moving the other plunger.

Fig. 206 shows the wiring diagram of this puzzle. This diagram is the bottom view of the chassis. All the wires which cross are not touching each other electrically. Insulated wire should be used and, if any bare wire sections touch another bare wire section, tape should be wound around the crossover or touching point to prevent shorting the electric circuit.

For the puzzle shown, $\frac{1}{2}$-inch plywood was used for the top panel of the chassis. The panel measures approximately 4 by 8 inches. The switches are the plunger type and reference should be made to Chapter 1 for a description of the construction details of the switch. A single battery is used as shown in Fig. 206, and holders for this battery can also be made from tin. The No. 112 flashlight bulb is mounted at the top panel.

There are only certain moves which can be made to prevent the bulb from lighting by the progress of the plungers up the board. An inspection of the wiring diagram shown in Fig. 206 will indicate what occurs. For instance, if the plunger in No. 6 hole is moved to No. 7, it will complete the circuit through the plunger remaining in the No. 1 hole and the bulb will light. Hence, the first move *must* consist of moving the left plunger from No. 1 hole to No. 2. After that, the plunger in the No. 6 hole can be moved to No. 7, etc. The sequence of moves which will solve the puzzle without causing the bulb to light is:

1 to 2
6 " 7
7 " 8
2 " 3
3 " 4
8 " 9
4 " 5
9 " 10

For grownups, the puzzle can be made even more complex by stipulating that after the plungers have reached No. 5 and No. 10 holes, respectively, they must be brought back to their original positions by moving the plungers one at a time without causing the bulb to light.

The puzzle would have to be worked a number of times before the proper sequence of moves is memorized; hence, it offers continuous diversion for some time.

After the constructor has become familiar with the wiring and operational details of the puzzle, a rearrangement of the wires can be undertaken to change the sequence of proper moves. The wiring can be modified to be just the reverse of what is shown in Fig. 206; that is, group No. 1 to 5 could be wired like group 6 to 10 and the latter wired like group 1 to 5.

Two persons can play the game, each selecting a plunger and deciding who moves first. Each player then makes one move at a time, but must move back when the light goes on. One point is lost when a move backward is made and, if one chooses to omit a move to permit the other to move more than once, he forfeits an additional point. When each player has reached the top of the board, he acquires 10 points less the deducted points.

advanced puzzles

ALTHOUGH the puzzles and games described here are more elaborate than those in the preceding chapter, they still use easily acquired batteries and light bulbs and easily constructed switches. These puzzles and games can be modified to meet the ideas and wishes of the constructor. It is always better, however, to build the puzzle or game first, make certain that it works properly, and then to make it even more elaborate, if so desired.

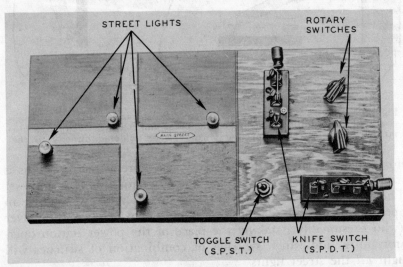

Fig. 301. *This is a more advanced type of switch puzzle. Using a variety of switches makes the game more interesting.*

Street-light puzzle

A more advanced version of the switch type puzzle is shown in Fig. 301. Puzzles of this kind prove more interesting when some reason is given for throwing the switches or some story is related to the solving of the puzzle. Also, if the puzzle is decorated in bright colors and such decorations are related to the particular

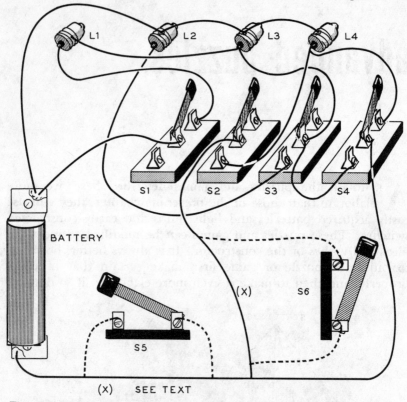

Fig. 302-A. *Wiring diagram of the Street-Light puzzle. Follow the wiring connections carefully and then place the bulbs and switches in any arrangement you wish. The connecting wires can be soldered directly to the light bulbs.*

story involved, the puzzle becomes even more attractive. This is true of the Street-Light puzzle. The story for this is as follows: The power station which lights the street lights is without an operator because the latter is away on a trip. Hence, the player (who presumably is taking the place of the power station operator) must find which switches or combination of switches will light *all* the street lights.

Four street lights are shown though more can be employed in

this puzzle. For the unit in Fig. 301 five switches are used though, as with previous switch puzzles, any number or variety of switches can be employed, hence the construction can consist of using whatever parts are desired or available. All the switches could,

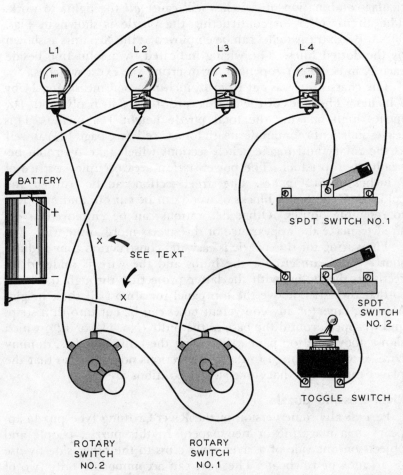

Fig. 302-B. *This is another version of the Street-Light puzzle showing how a variety of switches can be used.*

of course, be alike but the power-station section will appear more realistic and appealing to the player if the switches are dissimilar. Switches used for the puzzle shown consist of two knife switches, each a single-pole–double-throw, plus a toggle switch and two rotary switches. The toggle switch is a single-pole–single-throw device, while the rotary switches each have three positions. Actually, all the switching possibilities of the knife and wafer switches were not utilized, hence some dummy positions are present.

The wiring of the puzzle should be such that certain switch positions will cause two or three lights to glow. Only one particular combination of switches will cause *all* the lights to work. The circuit used for constructing the puzzle is shown in Figs. 302-A,-B. Extra switches can be employed at the X points as shown by the dotted lines. The wiring indicated by a solid line beside each X must be cut (opened) for inserting the extra switches.

The chassis top was cut from ¼-inch plywood measuring 14 by 7 inches. The sides of the chassis consist of ½-inch plywood, 1¼ inches high, making the total puzzle height 1½ inches. This leaves sufficient room underneath for a C cell (see page 124) as well as the rotary and toggle-switch sections which take up room beneath the top panel. The power-station area occupies a space of 7 inches by 5½ inches. The street section can be painted any colors desired. Small blocks of wood can be cut out and mounted to resemble houses. Other decorations can be employed as desired to make the appearance of the streets more authentic.

The wiring for this puzzle is easy and construction time is quite short. Holes are cut for the bulbs and the wire is soldered directly to the bulbs, with the latter projecting through the holes so that they stick above the top panel for almost all their height. Masking tape (or any equivalent tape) can be cut into thin strips and wrapped around the base of the bulb (No. 112 or 123) which shows above the top panel. In wiring the knife switches, dummy wires should be bolted to the knife sections not in use so that the player cannot tell that some switch positions are false.

River-crossing puzzle

Periodically some version of the River-Crossing type puzzle appears in a magazine or newspaper. In this puzzle, people and objects on one side of a river must cross to the other side by use of a canoe or rowboat. The boat can accommodate only two of the objects at one time, and the puzzle solver must try to determine how to transport the objects across the river without making any false moves.

A popular version of the River-Crossing puzzle involves a farmer who has in his possession a dog, a rabbit and carrots. He can transport only one object at a time across the river. He dare not leave the rabbit and the carrots on one shore while transporting the dog nor can he leave the dog and the rabbit alone while transporting the carrots. In another version the farmer has a goat, a wolf and some cabbage. He can't leave the wolf alone with the goat nor the goat alone with the cabbage.

Since the advent of electronics, the River-Crossing puzzle has reappeared wired with switches and indicating bulbs (or buzzers) and has proved of considerable interest. Toggle and knife switches, however, do not convey very well the idea of crossing a river, and

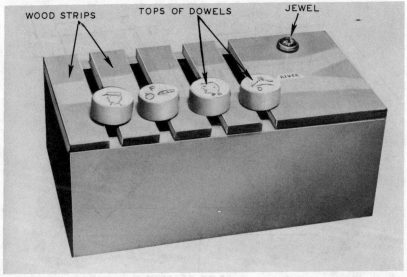

Fig. 303. *River-Crossing puzzle. The disks on top of the wooden dowels can be painted some identifying color or you can draw the characters they represent. A buzzer can be used in addition to the light shown in the photo.*

hence the version shown in Fig. 303 was devised. Here, the various objects are actually moved across the "river" by sliding the dowel section across the slots in the top panel of the puzzle chassis.

In this puzzle the characters have been changed in the interests of variety. Here, we have a Canadian mounted policeman who finds it necessary to transport a prisoner over a considerable distance to headquarters. Besides the prisoner, he has a bag of food and a dog with him. Again, as with the farmer, he must cross a river and the rowboat will only accommodate him and one other person or object. Hence, he must transport the prisoner, the dog and the food across the river (one item at a time) without leaving on either shore the food and the dog or the food and the prisoner. If he leaves the food and the dog, the dog may eat the food; and if he leaves the prisoner with the food, the prisoner would eat the food or escape with it. Because the river crossing is at a great distance from civilization, the prisoner would not

31

escape without the opportunity of taking the food with him.

The chassis for this puzzle measures 9 inches by 6, with sides 1 inch high. This height provides sufficient room for the particular switch contacts used as well as a battery and flashlight bulb. The height, however, can be increased if desired, as was done with the puzzle shown in Fig. 303, to accommodate the type of buzzer

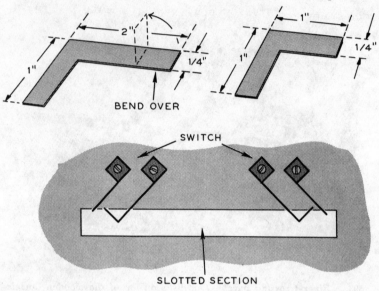

Fig. 304. *Switch construction details for the River-Crossing puzzle. The wooden dowel is inserted into the slotted section. The switch is closed by the movement of the dowel against it.*

described in the chapter on accessories. However, the 1-inch sides can be used initially, and higher sides made of 1/4-inch plywood added later if desired.

Four strips of 1/2-inch plywood are cut to measure 1 inch wide by 6 inches long. These are screwed or nailed to the chassis assembly as shown in Fig. 303, leaving enough room between the strips so that a wooden dowel or a metal tube will slide easily in the groove thus formed between the strips. For the puzzle shown, 1/4 inch was left between the strips. The spacing, however, depends on the type doweling or metal tubing available, but 1/4 inch or slightly larger is recommended because of better switch action.

The switch mechanism is shown in Fig. 304 and consists of tin strips mounted on the wooden panel sections. Note that one tin strip is bent out to be in position between the wood sections. Be-

cause of its location in the groove path, the movement of the
dowel in the groove will open or close the switches depending on
whether the dowel moves to the switch or away from it. An
actual mounting of the switch for one groove is illustrated in
Fig. 305. This must be repeated for the other three grooves.

Each switch is formed by cutting tin strips ¼ inch wide as fol-

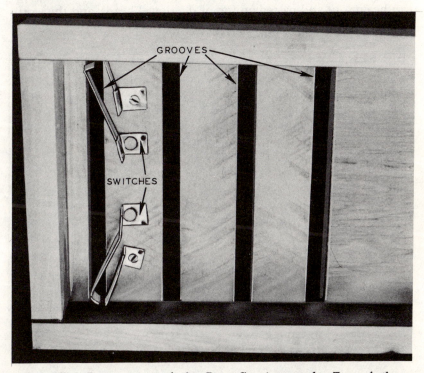

Fig. 305. *Bottom view of the River-Crossing puzzle. Two of the
switches are shown mounted in their correct positions. The wooden
dowel will slide along the vertical grooves shown in the photo.*

lows: three 1-inch-long sections and one 2-inch-long section. Two
of the 1-inch sections are soldered together to form an L as shown
in Fig. 304. The other 1-inch section is soldered to the end of
the 2-inch section as shown, forming a long L section. A 1-inch
section of each L is bent over to form an angle-iron mount. The
two sections are then screwed (or nailed with carpet tacks) to the
wood strips. The long section is bent out as shown, and the small
section is also bent so that no contact is made. When, however,
the dowel or tubing representing the "food" or "dog" is moved
along the slot between the wood strips, it will push the long tin
section closed, to make a switch contact.

The wiring of the puzzle is shown in Fig. 306 and represents the chassis *underside*. All wiring must be laid along the side panels so as not to interfere with the free movement of the dowel sections. The latter are cut to a length of approximately 1 inch, and a circular piece of wood is screwed to the top to act as a stop so that the dowel will not fall through the slots. Colored paper can be cut out and pasted on top of the dowel heads, and the words

BOTTOM VIEW

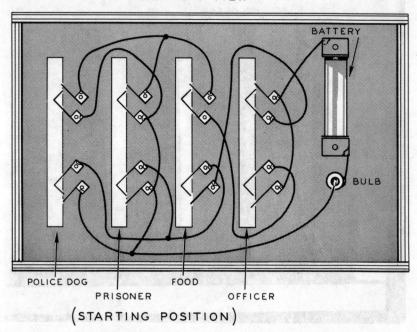

(STARTING POSITION)

Fig. 306. *Complete wiring diagram of the River-Crossing puzzle. The wires must be placed so that they do not interfere with the free movement of the wooden dowels when these are placed in their slots. Make the wires long enough so that they can be pushed out of the way if necessary.*

"policeman," "dog," etc., can then be written on them. A much better idea is to paste on pictures or drawings of a dog, food, a policeman, etc.

The flashlight bulb is for indicating a false move—the dog is left with the food or the prisoner with the food. When this bulb lights, all objects should be moved to their original positions and the player starts over.

Only a certain sequence of moves will permit the policeman to transport his prisoner, dog and food across the river. He must not leave on either shore the food and dog alone or the prisoner and food alone. Thus, the object of the puzzle is to move the policeman and *one* of the other items across the river. The policeman is allowed to go across or come back alone as required.

The proper sequence of moves involves the following:

1. Policeman takes food across.
2.   "   comes back alone.
3.   "   takes dog across.
4.   "   brings back food.
5.   "   takes prisoner across.
6.   "   comes back alone.
7.   "   takes food across.

Another solution besides the one given is for the policeman to take the prisoner across instead of the dog in move No. 3, taking the dog across in move No. 5.

The little thinker

An interesting type of puzzle is that in which the player is pitted against the puzzle and where specific moves must be made by the player in order to win. Such a puzzle becomes, in essence, a game in which the player must out-think the puzzle device or "machine." One version of such a puzzle game is shown in Fig. 307 and consists of 13 pegs. The player has a choice of removing one, two or three pegs at a time. When he has done so, the machine will indicate how many pegs it wants removed. Whoever is left with the last peg loses.

Either the player or the "machine" can have the first move. If the player moves first, he removes one, two or three pegs, as desired. He then depresses the button at the bottom of the panel. One of the lights at the right (marked 1, 2 or 3) will then light up, indicating the machine's choice. The number of pegs requested by the machine are then removed. The player then removes his selection of pegs and again depresses the button to indicate that he has made his choice. This continues until either the player or the machine is left with only one peg.

When the light (L) at the lower left goes on, it indicates that the machine concedes the game and hence the player wins. In explaining this game to the player, it is mentioned that since the machine can "think" ahead, it can ascertain before the final move is made whether or not it loses, and hence this light may glow before all the pegs are removed, indicating that the player has made the proper moves which would enable him to win.

Actually, of course, the machine does not "think" any more than the giant computers can. All the processes are built into the machine in logical sequence so that appropriate lights will turn on when certain pegs are removed. It is, however, difficult to beat the machine unless several games have been played and the player figures out the correct sequence of pegs which must

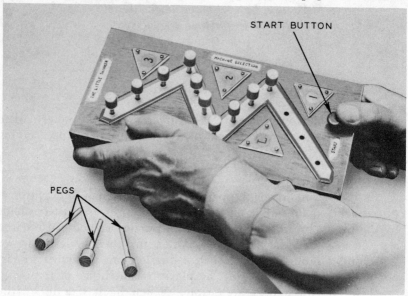

Fig. 307. *The Little Thinker is an unusual puzzle since the player matches his ability against that of the game. The puzzle can be arranged in a straight line or in the fashion shown in the photo.*

be removed. The trick, which of course should not be divulged to the player, is that he cannot win if he starts first. Regardless of whether he takes one, two or three pegs, the machine will beat him each time. If, on the other hand, the machine starts the game, the machine can still win if the player does not make the right moves. The machine's choice initially is always for the removal of one peg. If the player then removes either one or two pegs, he will not be able to win. If, however, he removes three pegs as his choice after the machine has indicated its first choice, then he has a chance to win if he "follows" through correctly.

The game shown was built on a chassis measuring 5½ by 10½ inches, though slightly larger dimensions can be used to avoid crowding of parts underneath. All wood used is ½-inch plywood and the side panels are cut 1½ inches high, making the total height of the puzzle 2 inches. As shown, the peg layout is in zig-

Fig. 308-A. *Another arrangement of the Little Thinker.*

zag fashion to break the monotony of having all the pegs in a straight line. However, the puzzle is much easier to build if the pegs are all laid out in a straight line, though in such an instance the chassis would have to be much longer than for the zig-zag version.

The flashlight bulbs are mounted in holes drilled into the chassis top and for the puzzle shown, triangular sections of plastic were used for covers. The plastic was sanded to make it opaque and also so that numbers could be drawn on the underside. If desired, however, the bulbs can be left uncovered and numbers marked at the side of each hole. Other coverings for the lights could also be used, such as jewels or reflector buttons. When such coverings are employed, the appropriate numbers can be shown beside each indicator light.

Plunger type switches are used for this puzzle but wood plungers are used instead of metal ones. These can be cut from wood dowels available at lumber yards or hardware stores. Use ¼-inch dowels for the size chassis indicated since this size works very well for the type switch to be employed. For the switches in this puzzle, a double function must be used. When the peg is removed, it must close two tin strips to make contact, at the same time opening another section. This is shown in Figs. 308-A,-B which give detailed drawings of the switch action, as well as the complete schematics for the puzzle. The bulb which lights when the "machine" concedes the game can be dispensed with if desired to cut construction time and to simplify the wiring a little. Without this light, the game is simply played to its conclusion, and whoever is left with the last peg loses. The "concede" light, however, is an added attraction which helps to give the illusion that the machine thinks ahead. If the concede light is omitted, connect X wire to point X at light No. 1.

The Little Thinker puzzle can also be constructed using metal-plunger switches. When this is done, however, the extra-contact tin strip must be bent around so that it makes contact *when the plunger is removed*. The complete wiring diagram is shown in Figs. 309-A,-B. In this puzzle, the "concede" light is omitted, but can be added if desired.

The Little Thinker puzzle can also be built using only 9 pegs, but the puzzle will have more interest if more pegs are employed. Any number can be utilized, as long as they are multiples of four plus one. Thus, a puzzle such as this could be built using 17 pegs, 21 pegs, 25 pegs, etc. A 21-peg version is described next, utilizing a different hookup principle.

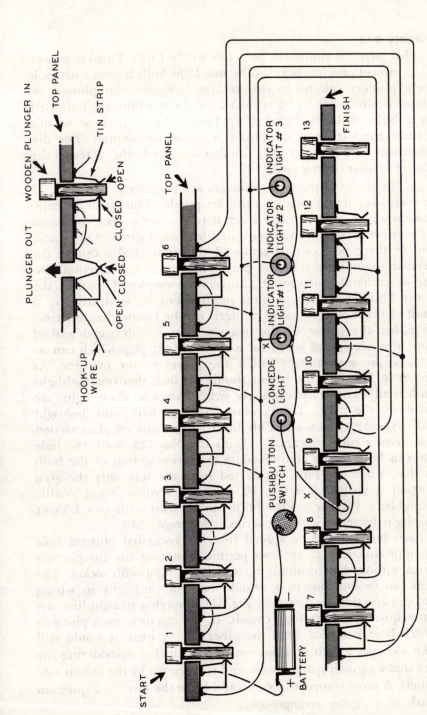

Fig. 308-B. *Side view and wiring diagram of the Little Thinker. The drawing at the upper right shows the action of the plunger.*

39

Twenty-one

This puzzle is similar in principle to the Little Thinker except that, instead of a few light bulbs, one light bulb is used with each metal-plunger switch. In this version, the particular plunger or plungers which are to be removed for the machine are indicated by a bulb or bulbs lighting up beside them. Because of this feature, the wiring of the puzzle is somewhat simpler. The disadvantage is that the greater number of light bulbs increases the cost of construction a little.

As the name implies, 21 plungers are employed. However, 17 or even 13 could be used for a simpler puzzle. The use of Twenty-one, however, lends greater interest to the game since the name "Twenty-one" has been popularized by a card game. Again, one, two or three pegs can be removed at a time, and a button depressed to indicate that the player has made his selection. On the other hand, since more plungers are employed than in the Little-Thinker version, a limit of four can be set instead of a limit of three on how many plungers can be removed at any time by either the player or the machine. For the device described here, the circuit was wired for a limit of four pegs which can be removed at any time by either the player or the machine. As shown in Fig. 310, the layout should be such that one flashlight bulb is adjacent to each plunger switch. Since so many lights are involved, the plastic covers were dispensed with, and flashlight bulb type 123 was used. This bulb has a rounded glass section and gives a nice appearance. With the No. 123 bulb, the hole size can be made to accommodate the screw section of the bulb so that, when the bulb is inserted from the top, only the glass portion is visible. If a No. 123 bulb (1.5 volts) is not readily available, a PR4 or a PR6 bulb can be used with two 1.5-volt flashlight batteries wired in series. (See page 124.)

Each bulb should be spaced from its associated plunger hole by approximately ½ inch to permit mounting the tin plunger strips without having the latter touch the light-bulb socket. The pegs can be laid out in a straight line for simplicity in wiring and construction, as shown in Fig. 311, though a straight-line layout requires a rather long chassis. If, for instance, each plunger-switch hole is spaced from the others by ¾ inch, it would still take a chassis length of approximately 17 inches, considering the fact that a ½-inch space at each end is taken up by the chassis side panels. A more compact layout could be in the form of a question mark or a zig-zag arrangement.

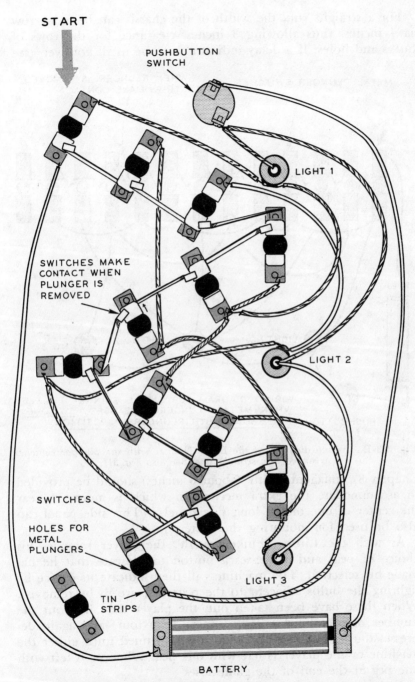

START

PUSHBUTTON
SWITCH

LIGHT 1

SWITCHES MAKE
CONTACT WHEN
PLUNGER IS
REMOVED

LIGHT 2

SWITCHES

HOLES FOR
METAL
PLUNGERS

LIGHT 3

TIN
STRIPS

BATTERY

Fig. 309-A. *Wiring diagram for the Little Thinker when metal plungers are used. See also Fig. 311.*

41

For a straight run, the width of the chassis can be as narrow as 4 inches, thus allowing 3 inches' clearance for the rows of bulbs and holes. If a delay indicator is to be used, however (see

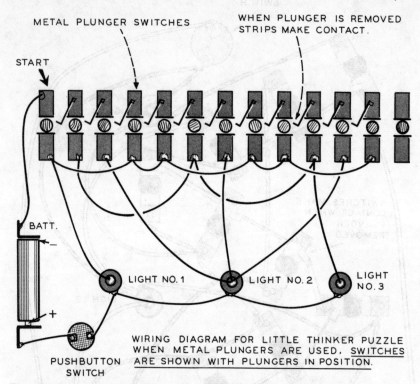

Fig. 309-B. *Wiring diagram of the Little Thinker with the plunger switches arranged in a straight line. See also Fig. 311.*

Chapter 8), sufficient width (about 6 inches) should be provided to accommodate the extra mechanism which is mounted near the center of one of the long side panels. (This side panel can also be used for mounting the battery.)

As with the Little Thinker puzzle, the player removes his choice of pegs and depresses a button to indicate that he has made his selection. The machine will then indicate its choice by lighting the bulbs adjacent to the pegs that are to be removed. When these have been taken out, the player then lifts out the number of pegs he desires (no more than four) and again depresses the button. This procedure is continued until either the machine or the player is left with one peg. Whoever is left with one peg at the end of the game loses.

To keep the game as simple as possible, no "give-up" light has been wired to show when the machine concedes the game. Instead, the game continues until either the machine or the player loses. Either the player or the machine can start first, and the circuit wiring automatically takes care of either sequence of moves.

As shown in Fig. 310 one side of all the light bulbs connects

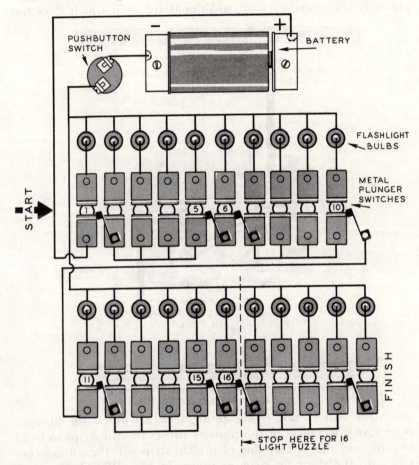

Fig. 310. *Wiring diagram for Twenty-One. This game is similar to the Little Thinker. Metal-plunger switches are used with light bulbs connected to each one. A smaller number of switches and bulbs can be used to make the game simpler.*

to the pushbutton switch which is depressed when a move has been made by the player. The other terminals of the light bulbs are wired to the respective tin strips of each plunger switch as shown. At the first plunger switch, one side of the battery is attached (plus terminal shown in the drawing). The negative terminal of the battery is connected to the other side of the pushbutton indicator switch.

The only plunger switches which require a little extra care in construction are Nos. 1, 5, 6, 10, 11, 15 and 16. These, in addition to the two tin strips for making contact with the metal plunger, also must have an additional tin strip which does not

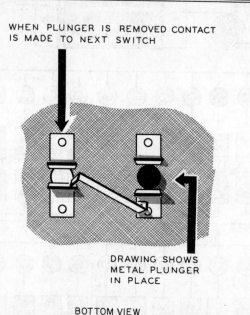

WHEN PLUNGER IS REMOVED CONTACT IS MADE TO NEXT SWITCH

DRAWING SHOWS METAL PLUNGER IN PLACE

BOTTOM VIEW

Fig. 311. *Bottom view of the metal plunger switches used in the game of Twenty-One.*

touch the battery side of the plunger switch unless the plunger is removed. Removal of the plunger causes the tin strips to bend inward, and the battery side of the tin strip will then make contact with the third tin strip, as shown in Fig. 310. The succeeding four plunger-switch sections are then connected in progressive order as shown. Fig. 311 shows how the plunger switches work.

If a shorter version of this game is desired, the plunger-switch sections can be wired only as far as the dotted-line section shown in Fig. 310. Hence, a 16-light unit can be built using the same principle. For those who are overly ambitious and wish to build a really lengthy version, as many additional plunger switches and bulbs can be employed as desired, by following the general sequence of wiring shown in Fig. 310.

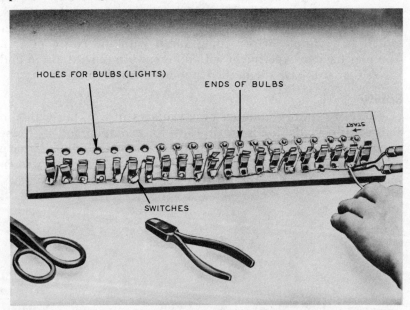

Fig. 312. *The game of Twenty-One can be wired as a single strip as shown in the photo. Holes are drilled in the wooden panel for mounting the light bulbs. Additional switches and bulbs can be added to make the game more complicated.*

The photograph shown in Fig. 312 illustrates how the game of Twenty-One can be wired as a single strip. Single-strip arrangements are somewhat easier to build and put in good working order. With experience, the plunger switches can be placed in staggered form, thus lending more interest and excitement to playing the game.

Working with plunger switches

The plunger switches used in these games can be made from scraps of tin. The action of the switch is such that the insertion or the removal of the plunger, whether metal or wood, closes and opens various circuits. This means that two actions take place at the same time. For example, putting a plunger into position may separate a pair of metal strips (thus opening a circuit) and at the same time one of these metal strips is forced into contact with an adjacent bit of tin (thus closing a circuit).

This opening and closing action means that the tin strips must

45

have a certain amount of springiness. A tin strip, when pushed aside by a plunger, must go back to its original position when the plunger is removed. If it does not do so, the game will not work properly. For this reason it would be well to build a single plunger switch as a sample before going ahead with the entire project. You may have to experiment with different bits of scrap tin to find pieces that will have the needed amount of springiness.

Safety

There are quite a number of advantages in working with tin. It is readily available (from tin cans, for example) and it is easy to cut with a pair of snips. The tin strips, however, will have sharp edges and can produce a severe cut. Also, be alert for burrs when pushing tacks through the tin strips or when using wood screws.

games for two players

GAMES and puzzles can be designed so that the player pits his skill against them, or against an opposing player. The games described in this chapter are intended for two persons playing against each other.

Hide-and-seek game

The plunger type switch described earlier is readily adaptable to the construction of games such as the one illustrated in Fig. 401. It is called Hide and Seek. An elongated chassis is used, with a partition in the center. It is intended for two players, each of whom has available six holes on one side of the partition. Each player is furnished with one metal plunger. The object of the game is for each player to seek out the metal-plunger position of the other player.

Each participant also has an indicator light on his side which will light when the position of his metal plunger coincides with that of his opponent's. The players decide who will go first, and each player, during his turn, is permitted two tries in seeking out the position of the opponent's plunger peg. Thus, if player No. 1 has the first turn, he inserts a plunger peg into any one of the six holes available on his side of the upright panel. The panel hides his position from his opponent. Player No. 1 then walks over to the other side of the board and watches while his opponent takes two turns in placing the plunger in two holes. If the bulb does not light, he has not made a "hit." Player No. 2 now takes his turn and places his peg in one of the six holes on his side while player No. 1 sits before his own panel section so that player No. 1 cannot see where player No. 2 places his peg. This routine is continued 10 times; that is, each has 10 turns. The

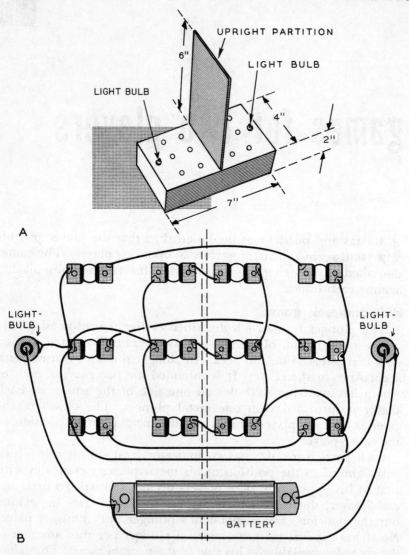

Fig. 401. *The upper drawing (A) shows the physical layout of the Hide-and-Seek game. The lower illustration (B) is the wiring diagram.*

one having the most number of hits during this time wins the game.

The suggested panel layout as well as the approximate dimensions are shown at A of Fig. 401, and the wiring for the puzzle is shown at B. The unmarked circles indicate the holes of the plunger switch, and the parallel lines on each side of the holes indicate the tin strips for the metal-plunger switch. The tin strips are bent so that when the metal plunger is inserted it will contact both of the metal strips and close the electric circuit for that particular hole. Make sure the strips are not bent together too much because there should be a separation between them when the metal plunger is removed.

This particular game lends itself readily to modifications. More holes can be added on each side of the upright panel and, if desired, each player can be assigned two plunger-switch pegs. Also, each player may be allotted three or more tries for each turn, if more switch positions are available on each side.

Trip to the shore

A more elaborate game using the plunger type switch is shown in Fig. 402. This game is also for two players. The sequence of plunger-switch holes forms a "route" and each player has his own route of travel, using a plunger which he moves from one hole to another as the game progresses. The number of holes which a player travels along the route is designated by an indicating "spinner." The object of the game is for one player to end at the finish line before his opponent. The metal-plunger type of switch is again utilized (see Chapter 1).

This game is of particular interest because no two games played are alike. The variation in plays comes about because the wiring of the game is such that the respective plays "change" and depend on the particular position of both players' plungers. Such variety is introduced by four indicator bulbs on the top panel which light up as various moves are made. The indicator light at the bottom is the "garage" and, when this turns on, the player's peg must be inserted next to the indicator light in that block. The path out of the garage is then shown and, as the dial is spun for advancing moves, the peg is moved out of the garage area and back into the route of the race. When the second indicator bulb lights, the player must move his peg back two spaces. Similarly, the third indicator light means a movement back of three holes. When the fourth indicator bulb lights, the player advances his peg five holes. Each hole can be considered a mile in the race.

The game is so wired that a particular indicator bulb will light only for certain positions of *both* pegs. Thus, if one player advances to hole 10 and an indicator bulb lights, that same bulb would not light when a player reaches the No. 10 hole while the other player was at some other position. The circuit for this game has cross-wires from one player's route to the other so that the indicator bulbs will light for different hole positions for each

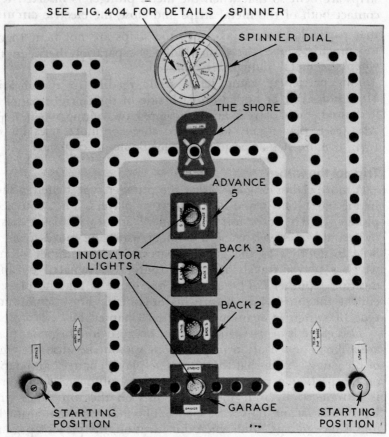

Fig. 402. *This photo shows a top view of the Trip to the Shore. Although it uses simple plunger switches, it is fairly elaborate and a large variety of games can be played. The players start at the lower left and right as indicated.*

game. This lends variety to the game and hence play is more flexible than would be the case if, for instance, a peg in hole 10 would always light the garage indicator.

The game is built on a chassis measuring 20 inches wide by 2½ inches high. The chassis is made from ½-inch plywood, a 20 by 20-inch panel of wood for the top and four 2-inch strips for the sides. When the strips are fastened to the top, the total height (including the top) will be 2½ inches.

The route layout should be somewhat as shown in Fig. 402 so that approximately the same number of corners exist. The reason for the sharp-turn corners is that the spinning indicator has one section on it which permits the player to advance to the nearest corner. Hence, some advances are only one or two holes while other advances may be as many as ten holes. The diameters of the holes should be such that the plungers which are to be used will fit into them easily without binding, yet without being too loose. For the puzzle shown, $3/8$-inch holes were used and each hole was spaced from the other by 1 inch to the nearest hole center. The flat portion of the plunger is bent over and a wooden knob fastened to it for convenience in moving the pegs along the route and inserting them into the various holes. Each peg can be colored differently from the other although, since each player has a route of his own identical to the other player's, an individual peg will be confined to a particular route and need not be identified from the other with respect to color. Coloring the pegs, however, helps dress up the game. Similarly, the route of each race can be painted a different color from the background to make it stand out and make the path more obvious.

There is a total of 48 holes for each route, plus the finish-line hole and the four holes necessary for the indicator bulbs. For the puzzle shown, the flashlight bulbs were inserted up through the indicator-light reflector button. These are available from auto supply houses and come in various colors. (See Fig. 809 on page 116.) Instead of using the jewel caps for the bulbs, the bulbs can project up through the panel without any caps, or colored cellophane can be cut into a small square and glued over the indicator-bulb holes.

If a drill press is available, the drilling of the various holes will be expedited. However, a hand drill can be used, taking care to make a clean cut through the panel. For the sake of appearance, the holes should be drilled from the top to prevent ragged holes such as may occur when the drill cuts through the wood. Usually, the end of the drill, as it pushes through the wood, will splinter it unless the drilling is done with an additional board underneath the hole being worked on. Since the bottom of the panel will not be visible, the edges of the holes can be sandpapered to remove splintered sections. The unevenness of the wood around the holes will be immaterial for the purposes of wiring.

A single battery in conjunction with No. 112 bulbs can be

used though, for longer-lasting battery-life, two standard dry cells can be hooked up in parallel as shown by the dotted lines in the wiring diagram.

After the holes have been drilled and before attaching the sides of the chassis, the strips of tin forming the plunger switches should be installed. Use short screws or carpet tacks.

The wiring diagram for the game is shown in Fig. 403. This arrangement has worked satisfactorily after some trial-and-error wiring and testing the game by playing it a number of times. The number of holes for each route does not have to be exactly 47, but could be less or more depending on the design features which the builder may wish to incorporate into the game. A fewer number of holes would, of course, cut down on the amount of construction required but it would also shorten the playing time. If room is available, the number of holes could be increased to 60 or 70. The playing of the version shown averages approximately 10 minutes, depending on how a particular game progresses. When the game is played, the indicator bulbs may light quite frequently and, by causing the pegs to be moved back or to the garage, would prolong the game. On the other hand, the indicator bulbs may not go on as often and a shorter game would result.

Once all the switch contact strips have been installed, they are wired as shown. You will note that, while there are a number of switches, their wiring follows a sequential order. The wiring time can be expedited if a single-length wire is soldered from one switch to another rather than cutting individual pieces of wire to connect one switch to the other. Wiring time is also shortened if wire is used which has no insulation. Bare wire of this type is used to hang pictures and can be purchased in 5-and-10¢ stores. The wire can be either solid or stranded but, in either case, the tinned type is preferred because it expedites the soldering process. When bare wire is used, care must be exercised not to cross two wires; crossed wires will cause a short circuit and would ruin the batteries. When it is necessary to place one wire over another, one of them should be wrapped with tape to supply insulation. As the wire is soldered to the switch contacts, route the wire around any of the holes and from the flexible section of the tin strips of the switch. By doing this, the flexing of the strip of tin

Fig. 403. *Wiring diagram of the Trip to the Shore. When inserting the two batteries, be sure they both face in the same direction. The plunger switches are shown in a straight arrangement to simplify the diagram.*

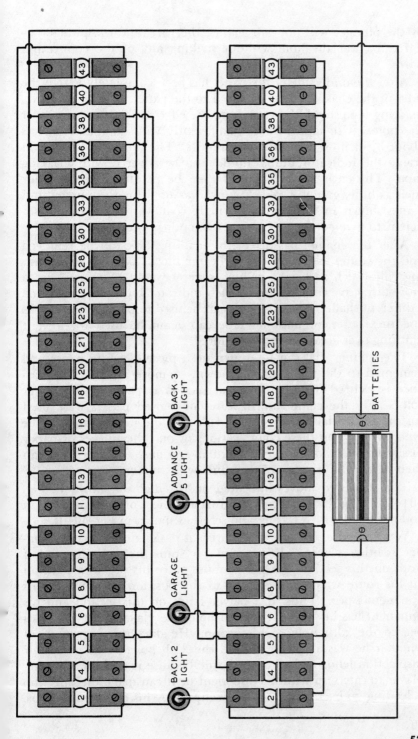

BACK 3
LIGHT

ADVANCE
5 LIGHT

GARAGE
LIGHT

BACK 2
LIGHT

BATTERIES

53

by the plunger will not rub against the wire, and the plunger itself can enter the hole without striking any of the connecting wires.

After the underside of the panel has been completely wired, the sides of the chassis can be attached to the panel. Before painting, masking tape should be run along the edge of the routes to cover the holes. Use tape which is sufficiently wide so that it covers about ¼ inch on each side of the holes. The square blocks which frame the indicator bulbs should also be covered with masking tape. The entire chassis should then be painted a single color such as blue, green or red. Enamel was used for the puzzles and games shown in this book. A single coat was found to be sufficient to give good coverage over smooth plywood.

After the enamel has dried, the masking tape can be removed and the routes as well as the indicator blocks painted a contrasting color such as bright yellow, silver or gold. You will get a more attractive coloring job if each indicator-bulb block is tinted a different shade. A small brush can be used to paint in the blocks and the routes, or masking tape can again be used to prevent painting over areas previously colored.

The spinner dial can be drawn on a piece of white paper and cemented to the top panel as shown. A more attractive appearance is secured if each triangular section is painted with water colors. For the game shown, different-colored papers were used for the triangular sections. The rules for the moves can then be typed directly on these color sections or on other pieces of paper dissimilar in color to the triangular sections. The sections can then be glued to the top of the panel with rubber cement.

The suggested moves indicated on the dial are shown in Fig. 404-A. The indicator arrow is cut from a piece of cardboard or tin and mounted with a washer and screw as shown in Fig. 404-B.

You will note that one of the areas is designated as "same number as other player." When a player spins the indicator and it stops on this section, the player must move his peg to a position on his route which corresponds to the position occupied by his opponent's peg. Thus, if a player lands on this section and his opponent has his peg in the 25th hole, the player must move *his* peg to the 25th hole of his group. He does this regardless of whether he was ahead or behind the 25th hole. This particular instruction lends variety to the game because it enables a lagging player to catch up with his opponent or it can make the one who is leading go back to a position occupied by his opponent. Thus, for the player who has not progressed far there is always hope of moving ahead to a position similar to his opponent's or of making his opponent move back to a position identical to his own.

After the game has been assembled and completed, it should be tested by referring to the wiring diagram of Fig. 403, and the pegs inserted into the appropriate holes which make the bulbs light. A check should also be made for the proper functioning of the cross-connecting circuits. For instance, if each player has

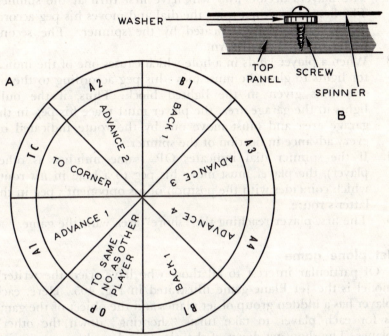

Fig. 404. *Drawing (A) shows how the spinner dial should be marked. Each section of the dial can be represented by a different color. The view illustrated in the upper right (B) shows how the indicator dial can be mounted.*

his peg in the fourth hole of his route, the garage indicator will light up. Recheck the garage-indicator light for holes No. 6 in each group. Then leave one peg in hole No. 6 in one route and place the other peg in hole No. 4 in the other route. The garage bulb should light up again. Also, place each plunger in hole No. 8 to light the garage. Leaving one plunger in hole No. 8, move the other plunger to hole No. 6 and then to hole No. 4 to make sure the garage indicator lights up for each of these moves. Follow this same procedure for testing the other indicator bulbs.

The game is made much more interesting by having a buzzer

sound for the advance light or for the other lights. Refer to Chapter 6 for constructional and wiring diagrams for the inclusion of a buzzer in this game.

Playing instructions—trip-to-the-shore game

1. Each player inserts his peg in the first hole at the start of his route.
2. The players decide who will have first turn at the spinner. The first player then spins the dial and moves his peg according to instructions indicated by the spinner. The second player then takes his turn.
3. When a player lands in a hole which causes one of the indicator lights to glow, he must move his peg according to the instructions given in the lighted block. Thus, if the bulb lights in the garage area, the player must place his peg in the garage area and must move out by the route indicated on every advance indication of the spinner.
4. If the spinner dial indicates OP (same number as other player), the player must move his peg to a hole in *his* route which coincides with the position of his opponent's peg in the latter's route.
5. The first player reaching the "shore" area wins the game.

Jet plane game

Of particular interest to all those who have tried the writer's model is the Jet Plane game illustrated in Fig. 405. Here, each player has a hidden group of jet planes and the object of the game is for each player to take turns "shooting" down the other's planes. The shooting down of the planes is done by a continuity-checking arrangement of the wiring.

The framework of the game consists of a vertical panel of 1/2-inch plywood, measuring 10 inches square. This panel is mounted on a chassis also 10 inches square and 2 1/2 inches high. The side panels are of 1/2-inch plywood cut 2 inches high so that, after the top panel is mounted, the total height is 2 1/2 inches. The vertical panel has a section cut out at the bottom center in the form of an arc so that a pushbutton switch can be mounted at the center for access by either player. An indicator light is mounted at the left of the flat portion of each player's side.

The holders for the tin jet planes are 36 metal rods which are mounted on the vertical panel. Each rod is inserted through the upright panel so that approximately 1/2 inch of the rod protrudes from each side of the panel. Thus, each metal rod establishes an electrical path through the upright panel.

Both players are allocated 10 strips of tin, each of which represents a jet plane. If small toy *metal* planes are available, they can be used, if holes are drilled in each wing for mounting on to the pegs of the upright panel. If tin strips are used, their ends must be bent so that they exert a slight spring tension when pressed between two of the pegs on the panel. Each player

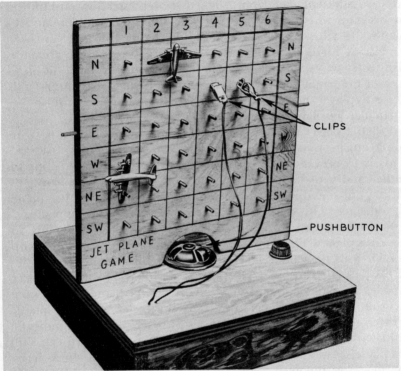

Fig. 405. *Photo of the Jet Plane game. The planes must be made of metal (do not use plastic) and can be obtained readily in toy shops. Small brass or tin strips can be used instead of the planes.*

mounts the jet planes across any 2 of the 36 metal rods or pegs of the upright panel.

The vertical sections of the panel are marked off in numbers from 1 to 6 for identification of the plane locations. The horizontal sections are marked as N, S, E, W, NE and SW. The horizontal designations represent points of the compass such as north, south, etc. The vertical numerical designations represent miles. When numbering the vertical sections, make sure that

No. 1 starts at the right on one panel and on the left on the other side of the panel so that the squares represented by the numbers will have the same numerical designation.

From each player's side two wires are available, at the ends of which are fastened alligator clips. (Any spring-loaded clip will do.) These two wires act as electrical feelers and aid the player in searching for the exact position of his opponent's planes. Thus, after the various planes have been mounted on the panel, the player having the first turn places the alligator clip leads on any two adjacent metal pins. He then depresses the pushbutton to ascertain whether or not he has made a hit on his opponent's planes. If a hit is made, the indicator lights will glow on both panels. A hit is recognized by the device because the alligator-clip feeler wires will make a complete circuit through the metal rods of the panel and through the tin jet plane on the other side. Once a player makes a hit, his opponent must relinquish the particular jet plane which has been struck.

When a player makes a hit as evidenced by the indicator lights going on, he must call out the location, for example, "W-4-3." This is to prevent a player from shorting the alligator clips together and thus falsely causing the indicator lights to light. By calling out the alphabetical and numerical location of the plane, he indicates to his opponent that an honest hit has been made.

Each player gets three turns at a time in his search for his opponent's planes. The one who first captures all the planes of his opponent wins the game. During a player's turn, he may move any of his jet planes to other positions as desired. This rule is necessary because his opponent may have some jet planes in the same position on the panel. Naturally, if the feeler wires are placed across pegs which contain a player's own jet plane, the indicator light will show a hit. Thus, the player during his turn can shift his plane to another position if he wishes to make a check across the two pegs involved. He can then return the jet plane to the original position or leave it in the new position.

During a player's turn, the feeler wires on the opponent's side must be separated so that no false indication is obtained. A good plan is to have two dummy pegs mounted into each side of the panel as shown in Fig. 406. The unused feeler wires of one player can then be placed across these dummy pegs to prevent false indications while his opponent takes his turn.

With each player having 10 metal strips representing jet planes, the game takes some time to complete because the search for the opponent's planes becomes more difficult when there are fewer

planes left to find. To shorten the game, it can be decided that
the game is over when one of the players captures, say, five of his
opponent's planes.

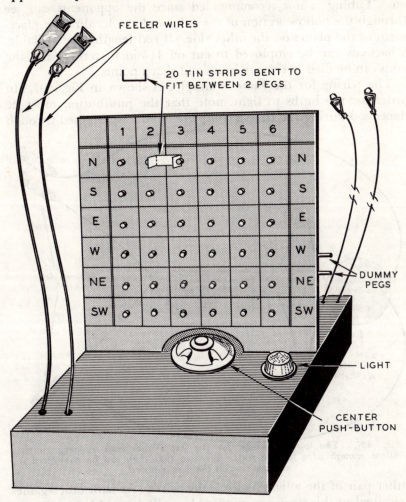

FEELER WIRES

20 TIN STRIPS BENT TO
FIT BETWEEN 2 PEGS

	1	2	3	4	5	6	
N	○	▭	○	○	○	○	N
S	○	○	○	○	○	○	S
E	○	○	○	○	○	○	E
W	○	○	○	○	○	○	W
NE	○	○	○	○	○	○	NE
SW	○	○	○	○	○	○	SW

DUMMY
PEGS

LIGHT

CENTER
PUSH-BUTTON

Fig. 406. *Further details of the Jet Plane game. The dummy pegs can
be used to hold the feeler wire clips when these are not being used.*

The vertical panel should be laid out initially as shown in Fig.
406. A ballpoint pen can be used for drawing the lines on the
panel and for lettering in the numbers and compass points. After
that, through the center of each square a hole is drilled of sufficient size to accommodate the metal rods so that they fit rather
snugly. The rods can consist of nails or screws having a length of

approximately 1½ inches. A better job results, however, if solid rod sections are cut from the center rod of a curtain-rod assembly or some other solid metal rod such as the ends of indoor antennas, etc. Tubing is not recommended since the opponents can see through the hollow section of the tube and can localize the placement of the planes on the other side. If rod lengths are available, a hacksaw can be employed to cut off 1½-inch sections and the ends can be filed to remove burrs and sharp points.

The wiring for the game is simple, as shown in Fig. 407. In order for the bulbs to light, note that the pushbutton must be depressed during the time a complete circuit is obtained through

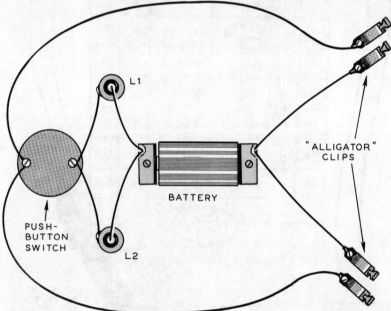

Fig. 407. *The wiring needed for the Jet Plane game is very simple. Allow enough wire for the feeler wires so that they can be connected to any position on the center board.*

either pair of the alligator-clip feeler wires. As with other games described in this book, an improved version would have a buzzer as well as a light. For buzzer details, see Chapter 8.

Playing instructions—jet plane game

1. Each player mounts his 10 metal strips (jet planes) across any two of the metal rods according to his choice.
2. The players decide who goes first. The first player connects electrical feelers across two of the pegs on his side and depresses the pushbutton to ascertain whether or not a hit has

been made. If the indicator bulb lights, his opponent must relinquish the particular jet plane which has been hit. Each player gets three turns at a time.

3. When a hit is made, the player making the hit must identify the position of the plane by the letters and numbers on the panel.

4. During any player's turn, he may move his own jet planes in order to "clear" two pegs so his own jet plane will not give a reading.

5. The first player to capture five planes (or any previously determined number) from his opponent wins the game.

The electrical "feeler" wires with the alligator clips attached to their ends should be flexible enough so that they can readily be placed in any desired position on the panel. Radio hookup wire having either a plastic or other type of insulation can be used. Wire size should be no larger than No. 18 or No. 20 (No. 20 preferred). Stranded wire should be used because of its flexibility. Solid wire is not as flexible and is likely to break after the game has been played a few times. With No. 20 stranded wire, with a thin insulation, good flexibility will be obtained and wire breakage will be postponed for some time.

If radio hookup wire is not readily available, the electrical feeler wires can be made from discarded line cords from old lamps, radios or other appliances. Use the type of ac line cord which has rubber or flexible plastic insulation. Cut off a length sufficient for the puzzle and carefully slit the wire at one end by using a razor blade to cut the wires apart for an inch or two. Make sure the razor blade cuts through the center of the insulation between the two wires. After a few inches have been split apart, the remainder of the cord can be separated by pulling the ends. When using discarded ac line cord in this fashion, it is advisable to make sure that the wire is unbroken within the insulation. Attach the two wires procured from the ac line cord to a flashlight bulb and connect a battery at their other ends. If the flashlight bulb lights, the wires are all right.

Alligator clips are available at dime stores, radio supply outlets, hardware stores and model-parts supply houses. The clips have a spring action so that the jaws open when pressure is applied. A connecting screw at one end facilitates attaching a wire to the clip.

The Jet Plane game with 36 squares has a fairly long playing time. If a quicker game is desired, use only 16 squares. In this, the horizontal top numbers will go from 1 to 4, and the vertical designations at the sides will consist of only North, South, East and West, omitting the Northeast and Southwest sections. When such a smaller version of the game is planned, the dimensions given in the text can be reduced accordingly. With the smaller number of squares, a fewer number of metal strips (jet planes) are allocated to each player. Four planes is a good number.

If desired, of course, a larger-sized game can be constructed by adding two more rows vertically as well as two more rows horizontally. The addition of such rows would provide a total of 64 squares. At the top, horizontally, numbers would run from 1 to 8, and vertically, at the sides, the additional compass designations can consist of Northwest (NW) and Southeast (SE). Again, the dimensions of the game must be changed to accommodate the increased number of squares. When the game squares are thus increased, it is not necessary to increase the allocation of metal strips to each player. Each player can still receive 10, but the first player to capture three planes would win. With the larger number of squares, seeking out planes is more difficult and game time is prolonged. Hence, for the larger game, it would be preferable to set as a limit the capture of three planes rather than five.

The Jet Plane game lends itself readily to variations in size because neither the wiring nor electrical connections of the game are affected. When increasing the size of the game, however, make sure the feeler wires are sufficiently long so that the alligator clips can be attached to any section of the vertical board, particularly at the upper right and left corners.

games for several players

Of necessity, games for groups are somewhat more complex than those that have been described earlier. These games present more of a challenge to the constructor since the interest of each contestant must be maintained until the game is completed. For that reason, it is well to consider the addition of bells, flashing lights, buzzers and other attention-compelling devices.

Cross-country race

The game illustrated in Fig. 501 is designed for four contestants, but it can be played by only two or three or as many as five or six. Each person is assigned an individual peg just as in the Trip-to-the-Shore game. The object of the game is to have each player move his peg along the route laid out from the starting line, each player going along the same route. The first to reach the finish line wins.

This game is interesting because it combines a number of variable and fixed indications. The variable indications are those which may or may not function, depending on the position of the various players. One such variable indication is the detour sign which compels a player to take a longer route if some of his opponents are lagging behind. The short cuts are also variable and will light up only if opponents are ahead of the player. In such an instance, if the player reaches a given point in the route, the short-cut light will go on, enabling him to catch up with his more advanced opponents. Stop signs are also variable; they will not light unless two players reach the "intersection" of the routes at the same time.

The fixed indications are the advance 4 and the back 3 lights. The actuating holes for these are located at several positions along the route and will light up whenever a peg is placed in the proper position. The speed court is a combined variable and fixed indication. The hole preceding the speed court is a variable indication because it will not light unless there are some lagging opponents in back of the player. The hole opposite the indicator and the hole beyond the indicator are fixed positions which will light the bulb whenever a plunger is inserted.

Two spinning type indicators are provided; one a high-speed and the other a low-speed unit. Each player, in taking his turn, has a choice of using either the high- or the low-speed spinner. The high-speed spinner permits a faster advance but also provides greater hazards in making a player go back a greater number than the low-speed spinner. On the high-speed spinner the player can advance as much as five holes or go back as much as three. Two go-back areas are present on the high-speed spinner. With the low-speed spinner, the highest advance is three but only one go-back number is present (B2). By providing two spinners, however, the game becomes one less of chance and more a matter of which spinner the player selects.

As with the Trip-to-the-Shore game, the entire game should be painted some primary color for the background, such as blue, red or green. The speed court and the advance and go-back areas should be painted a different color from the background. In the puzzle shown in Fig. 501, the triangular starting-position area was painted green. The route was colored yellow to make it distinctive from the remainder of the puzzle.

When two players reach an intersection and cause operation of the stop light, they cannot advance until it goes out. Since the light will not extinguish until one of these positions is unoccupied, it means that each stays there and takes his regular turn at the spinner dial. When one of the players must go back as ordered by the spinner indication, the stop light will go out. The other player may now advance if the spinner indicates an advance on his next play. It is possible that both may be compelled to go back and in coming forward both may reach the stop area again.

The speed-court section has provisions for the four players in case all should reach this area. When the speed-court indicator is lit, the peg must be removed from the particular hole it occupies and be placed in one of the holes of the speed-court area. The player must then wait here until it is his turn again, at which time he may advance out of the speed court and back into the regular traveling route if the spinner numbers permit him to do so. If, on an advance, a player finds the hole occupied by another player, he places his peg in the hole beyond the one occupied. On a go-back move, he must place his peg in a hole in back of any occupied hole.

The game is built on a chassis measuring 16 inches by 19 inches

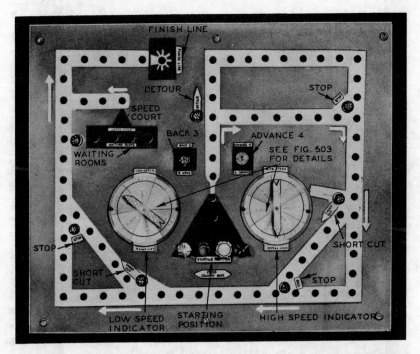

Fig. 501. *Top view of the Cross-Country Race. This game was designed to be played by a group. Two spinner dials are required, adding considerable variety. Metal type plungers are used, as in the games described earlier.*

by 2½ inches high. It is made of ½-inch plywood. The side panels are cut 2 inches high and in combination with a top provide a total height of 2½ inches as was the case with the Trip-to-the-Shore puzzle. The size of the holes depends on the diameter of the metal rods available. If ³⁄₁₆- or ¼-inch-diameter rods are employed, radio knobs can be used. This was done with the puzzle shown in Fig. 501. Two white and two black radio knobs were used, one white knob being circular and the other triangular. The black knobs are also distinguished from each other by the same difference in shape. If desired, the top of the metal

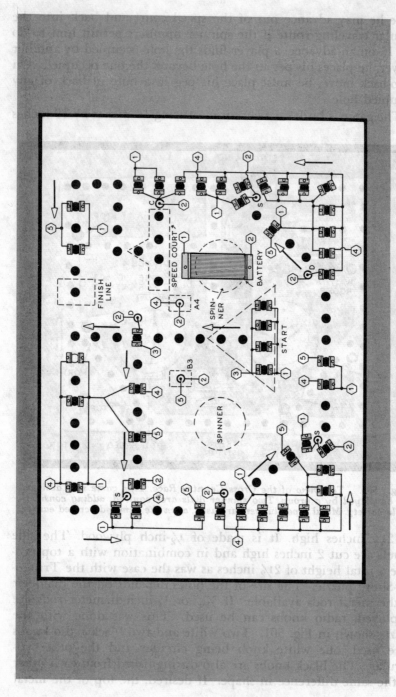

Fig. 502. *Wiring diagram of the Cross-Country Race. All points having identical numbers should be connected by wires. For example, all symbols marked 1 should be connected; then all symbols with the number 2 should be wired together, etc.*

plunger can be bent over and a piece of wood fastened to the top for a handle. Each piece of wood can then be painted a different color to distinguish one plunger from the other. Each plunger can also be numbered by painting a large-sized 1, 2, 3 or 4 on top of the flat section for easy identification as the pegs are moved along the route.

The route layout should be approximately as shown in Fig. 502 though a greater number of holes can be employed if a larger

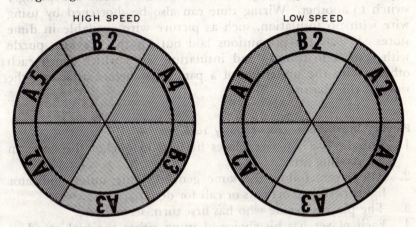

Fig. 503. *Here are suggested designs for the high-and-low-speed spinners. Identifying colors will make the spinners more attractive.*

puzzle is desired. With four players, however, time is consumed waiting for each one to take his turn and, consequently, the game playing time is not too short.

The indicating lights for the detour, short cut, stop and speed court were decorated by capping them with radio pilot-light jewels. These are small in size, will fit readily into a 7/16-inch hole and do not take up much room. Any larger bulb covering would prevent insertion of the plunger into nearby holes. If larger indicator decorations are utilized (such as the reflector buttons used on license plates) more room must be provided for the indicators, hence a larger puzzleboard would be needed. The license-plate reflector jewels can, however, be used on the advance and go-back sections.

As with the Trip-to-the-Shore puzzle, the indicators are made up of various-colored sections to add to their attractiveness. The suggested marking of the areas is as shown in Fig. 503. After the spinner dial has been glued to the top of the puzzle, circular sections can be cut from plastic or cellophane and either glued or screwed over the top of the indicator dial to prevent soiling when in play. Indicator arrows are then cut from tin and mounted as shown in Fig. 501.

The wiring diagram for the game is more simple than the Trip-to-the-Shore game because fewer variable sections are used and two separate routes are not utilized. The wiring of the switches follows a sequential order and the wiring time can be expedited if a single-length wire is soldered from one plunger switch to another. Wiring time can also be shortened by using wire with no insulation, such as picture wire available in dime stores. Follow the precautions laid out for the previous puzzle with respect to routing and insulating wires which touch each other. Wire the underside of a panel first before attaching the sides of the chassis.

Playing instructions—cross-country race

1. Initially, each player places his peg in one of the holes in the starting area.
2. All players follow the same general route unless indicator lights permit short cuts or call for detours.
3. The players decide who has first turn, second, etc.
4. Each player has his choice of using either the high-speed or low-speed spinner.
5. If the spinner indicates an advance and an opponent's peg is in the hole to which a player's peg must move, the player places his peg in the next hole beyond the one occupied. If the spinner indicates a go-back move and the hole is occupied by another peg, the player must place his peg in the hole in back of the one occupied.
6. When the stop sign lights, the players cannot advance until the light goes out. The light will go out only if one of the players goes back because of a spinner dial indication.
7. When a player occupies a hole which lights the speed-court indicator, he must place his peg in the speed-court area and wait his turn to advance out of this area and back into the regular route.
8. A player reaching a hole which lights either the go-back or advance indicator bulb must move his peg accordingly. As with the spinner indicator, on an advance move the peg goes ahead of any hole occupied and on a go-back move the peg goes in back of any hole occupied by another peg.
9. The first player reaching the end of the trip wins the game.

Space-travel game

Other games for several players can be constructed by varying the routes, the spinner indications and changing the central

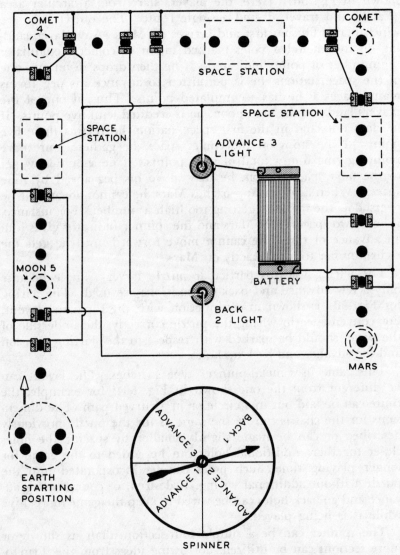

Fig. 504. *Wiring diagram for the Space-Travel game. For the spinner, follow the same instructions given for games described earlier.*

theme of the Cross-Country Race game. In the previous game the basic idea was a race across the country. Another type of game for two or more players can be built around a hurdle race, a golf course or any other idea which the constructor thinks would be interesting. A suggested variation is the Space-Travel game shown in Fig. 504. Here, the players start from a circular area (Earth) and travel around a single route. The object is to see which player finishes first and arrives on Mars. Any player landing on the encircled poles marked moon or comet accumulates the number of points indicated. If he then drops in on any one of the space stations, he is permitted to advance his peg for as many spaces as he has accumulated points. Thus, if one of the players lands on the first moon, he is credited with five points. If he does not land in the first space station, he retains these five points. If he stops on the comet space, he gathers four more points, giving him a total of nine points. If he should now get to the next space station, he may move his peg ahead by nine spaces. When a player approaches Mars, he cannot advance if he overshoots the mark by getting too high a number. For instance, if he is two spaces from Mars and the spinner indicator gives him an advance of three, he cannot move forward until he gets the right number to land exactly on Mars.

The wiring for this particular puzzle is very simple because only a few advance and back-up positions are used. The wiring for Fig. 504 is shown in conjunction with the top-panel layout. For convenience in wiring the puzzle correctly, the underside of the panel should be marked with respect to the starting position and the advance and back-up holes.

This game uses metal-plunger type switches. The layout can be different from the one shown in Fig. 504; for example, the route can be laid out in a circle or in a curved path. The dimensions for the chassis can be the same as for the puzzle previously described or can be made slightly smaller by spacing the holes closer together. Additional holes can be added to the game for longer playing time. Such holes can be incorporated into the puzzle without additional wiring. If desired, of course, extra advance and go-back holes can be wired to keep the game more alive while it is being played.

The spinner can be a simple four-section affair as shown or more sections can be utilized, borrowing ideas from the Trip-to-the-Shore and the Cross-Country Race games. Thus, if the route is laid out to provide more corners, one of the spinner sections could indicate an advance to the nearest corner. It is advisable, however, to avoid more than one back-up section on the spinner because, if the players have to go back too many times, they will lose their enthusiasm for playing the game. For that reason, too, it is better to keep the back-2 light holes to a minimum.

As with the Cross-Country Race, if one player is to advance to a certain hole and finds it occupied by another player, he may place his peg ahead of the occupied hole. If, on the other hand, he is compelled to retreat either because of the back-up light or the spinner indication, he must place his peg behind any peg occupied. For added interest, both the advance and back-up lights can be wired for a buzzer as described in Chapter 8.

Number of players

Five holes are shown at the starting position for the Space-Travel game. Thus, this game can be played by as many as five players.

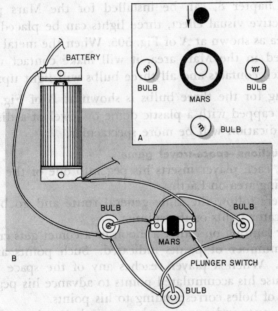

Fig. 505. *Three flashlight bulbs can be placed around the Mars area (A). The wiring is shown in the lower drawing (B).*

The game can, of course, be designed for more or fewer players as desired. If five is chosen as a limit for the number of players, then five metal plunger pegs of the type described in Chapter 1 must be constructed. The wooden mounts for the metal plungers can be identified by painting each a different color or numbering the pegs. Better identification is secured if each peg is a different color and also has a large number painted or drawn on the top of the mount.

As with other games, the top panel can be made more attractive by painting the various sections in different colors. The comet areas, for instance, can be painted in orange, Mars in red and Earth in green. Space stations can be painted brown, and the background of the panel can be painted blue to represent the sky.

Other holes can readily be incorporated if desired. Also, if desired, the end of the game can be indicated by one or more lights going on when the player inserts his plunger in the Mars area. A winning player usually likes a visual or audible indication that he has reached the end. A buzzer, constructed as described in Chapter 8, can be installed for the Mars position. For an attractive visual effect, three lights can be placed around the Mars area as shown at A of Fig. 505. When the metal plunger is then placed in the Mars area, it will make contact with the plunger switch contacts and all three bulbs will light up.

The wiring for the three bulbs is shown at B of Fig. 505. If each bulb is capped with a plastic dome or jewel of a distinctive color, the indication will be more spectacular.

Playing instructions—space-travel game

1. Initially, each player inserts his peg into one of the holes in the starting area on Earth.
2. All players follow the same general route and go by either the indicator lights or the spinner.
3. A player landing on either a moon or a comet gets credit for for the number of points indicated. Such points are accumulated. When a player reaches any of the space stations, he may use his accumulated points to advance his peg by the number of holes corresponding to his points.
4. If the spinner indicates an advance and the player finds the hole to which he is supposed to advance occupied, he places his peg in the next hole beyond. If the spinner indicates a go-back move and the hole is occupied by another peg, the player must place *his* peg in the hole in back of the one occupied.
5. The first player reaching the hole marked Mars wins the game.

miscellaneous puzzles

THE puzzles in this chapter (and the games in the next one) deviate somewhat from normal routine in one or more items, such as different switch types, unusual layouts, etc. This is no indication that they are more difficult to construct or more complex in their wiring; they simply present different ideas. Also, these particular puzzles and games are of such a nature that they lend themselves readily to such variations and changes as may be desired by the constructor. Hence these games and puzzles serve as examples of how a basic item can be altered in design to conform to the particular ideas of the individual constructor. By taking into consideration the manner in which a basic puzzle or game can be changed to increase its scope or be otherwise improved, the same underlying principles can be applied by the reader to other units or to originals.

Because the items in this chapter and the next are discussed with respect to various forms which such units can take, chassis or panel dimensions are not provided since the constructor can readily ascertain such dimensions by judging the size needed for the layout he chooses.

Mystery safe puzzle

The Mystery Safe puzzle shown in Fig. 601 employs an easily built rotary type switch. A strip of tin is used for the rotating or selecting member of the switch and large-size (unpainted) thumbtacks are employed for the contact points. This switch is a rotary multiple-contact type and the rotary metal strip must move over one of the thumbtacks to make contact.

As you will note in Fig. 601 there are three dials. These can be laid out as shown in the illustration or all three placed side by side. The thumbtacks of each dial are numbered consecutively from 1 to 8. The object is for the player to move the metal dial slides around until he hits the right combination to "open the safe." Three flashlight bulbs are mounted on the panel and, when all three are lit, the correct combination has been found and the safe has been "opened."

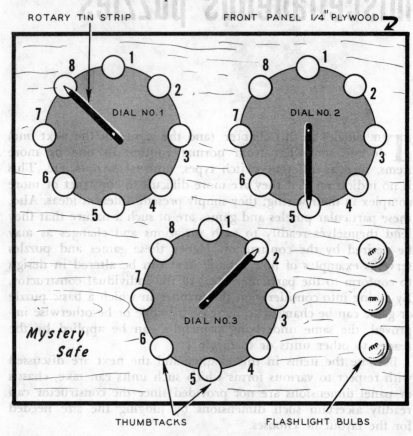

ROTARY TIN STRIP   FRONT PANEL 1/4" PLYWOOD

THUMBTACKS   FLASHLIGHT BULBS

Fig. 601. *Front-panel view of the Mystery Safe puzzle. Thumbtacks are used as the contact points.*

Fig. 602 shows a suggested wiring diagram for this puzzle. Note that the three flashlight bulbs are wired in parallel so that all three will light at one time. With this basic puzzle it is not necessary, of course, to use three flashlight bulbs, since one will suffice for an indication of when the puzzle has been solved. The use of three, however, gives a better visual effect. Depending on the wishes of the constructor, however, one, two or three can be used as desired.

The wiring shown in Fig. 602 is such that the proper combination to "open" the safe is 2 — 6 — 3. Only when the three dials are set on these numbers, respectively, will the flashlight bulbs operate.

The flashlight bulbs and the battery are mounted beneath the panel in the manner described for puzzles and games in earlier

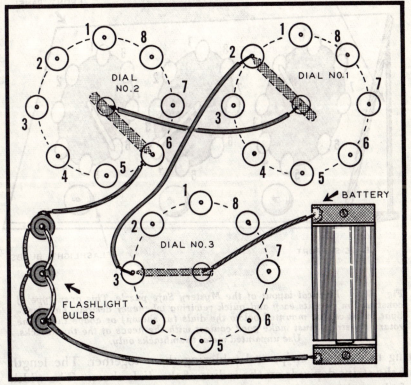

Fig. 602. *Wiring diagram of the Mystery Safe puzzle. Once the correct sequence of numbers is learned, the puzzle can be rewired to change the combination.*

chapters. Use 1/4-inch plywood for the dial panel because the thumbtack point will penetrate wood of this thickness and appear beneath it. Thus, the contact wire can be soldered easily to the thumbtack point which projects beneath the panel. The panel itself can be mounted on a square or rectangular boxlike frame in a fashion similar to that illustrated for previous puzzles and games. As a variation, however, it is possible to mount the panel vertically with a slight backward tilt as shown in Fig. 603. Here only front and base-support panels are needed. The sides and back are open. The battery can be mounted on the bottom panel or on the back of the front panel.

The sliding contact arm construction is shown in Fig. 604. The sliding arm should be of fairly stiff tin, such as that used for roofing. If tin from cans is used, double the thickness by cut-

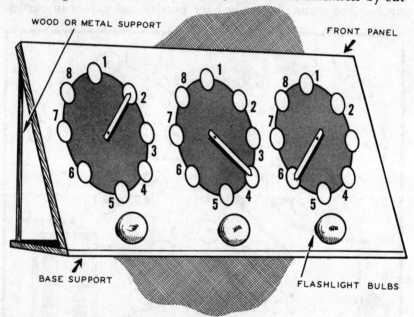

Fig. 603. *Physical layout of the Mystery Safe puzzle. This open type of construction permits easy and quick rewiring whenever desired. The flashlight bulbs can be mounted below the dials (as shown) or above them. The rotary tin strips must make good contact with the heads of the thumbtacks. Use unpainted metal thumbtacks only.*

ting two similar strips and soldering them together. The length of the strips depends on the radius of the dials chosen. The width of the strips can be 1/4 or 1/2 inch, depending on the stiffness desired. Taper the tip of the tin strip to form a blunt point. Bend the tip over slightly so it makes contact with the thumbtack and still clears the panel without binding.

A hole is drilled at the end of the tin strip, opposite the tapered end. A hole is also drilled through the center of the panel dial to accommodate a bolt as shown in Fig. 604. A washer separates the sliding arm from the panel so as to lift the arm slightly above the panel for free movement. Another washer is placed beneath the panel for soldering the connecting wire to the unit, hence this bottom washer should be cut from tin. A diameter of approximately ½ to 1 inch for this washer will be suitable. Punch or drill a hole through its center.

A spring is placed between the bottom holding nut and the tin washer to provide sufficient tension for good electrical contact. The lead wire is soldered to the tin washer and run to the respective terminals shown in Fig. 602. Another wire is soldered to the point of the thumbtack which protrudes beneath the panel (for those thumbtacks which are connected into the circuit). If

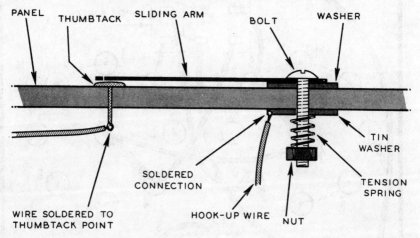

Fig. 604. *Side view of the construction of one of the dials of the Mystery Safe puzzle. The nut should be tightened so that the dial turns readily with finger pressure. If the nut is too tight, the dial will not turn.*

sufficient length of thumbtack point is not available, use the sharp end of a knife to cut away some of the surrounding wood until enough thumbtack point protrudes for soldering.

A variation of the basic Mystery Safe puzzle is shown in Fig. 605. A double-pole double-throw switch has been incorporated to change the combination when desired. With the switch thrown to the right position, the combination is 2 − 6 − 3 as for the puzzle originally shown. When, however, the switch is thrown to the left, the new combination is 4 − 8 − 3. If the open knife type double-pole double-throw switch is used, changes of combination in addition to the foregoing are easy because connections to the switch are made with setscrews. Several wires could be soldered to various thumbtacks and left unconnected from the switch until a change of combination is desired. For instance, a wire could be soldered to thumbtack No. 1 on dial 2, and at a later time the lead from thumbtack No. 1 could be connected to the double-pole double-throw switch in place of the wire from No. 8 thumbtack. Similarly, a lead could be attached to No. 5 thumbtack on dial 2 for connection in place of thumbtack No. 6. Similar "combination-changing" wires can be attached to the thumbtacks for dials 1 and 3.

Note that the three flashlight bulbs in Fig. 605 have been wired in *series* instead of in parallel as was done for the wiring

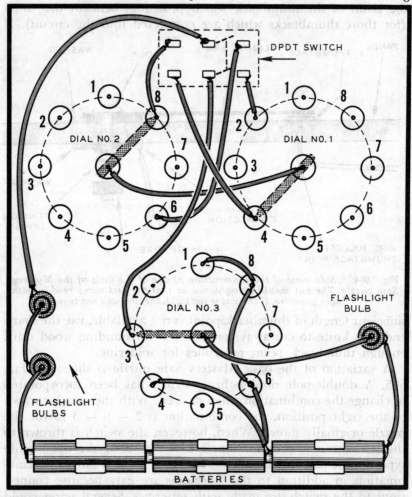

Fig. 605. *Variation of the Mystery Safe puzzle. The double-pole double-throw switch supplies a quick change of the combination. Three batteries are needed for this puzzle. These batteries are connected in series—that is, the negative terminal of the first battery touches the positive terminal of the second battery, etc.*

diagram of Fig. 602. Also, three batteries are used instead of one, to provide for the additional voltage required by the three series bulbs. (Three 1.5-volt bulbs in series require 4.5 volts. It is obtained by placing three 1.5-volt batteries in series.)

The series connections provides a variation with respect to dial

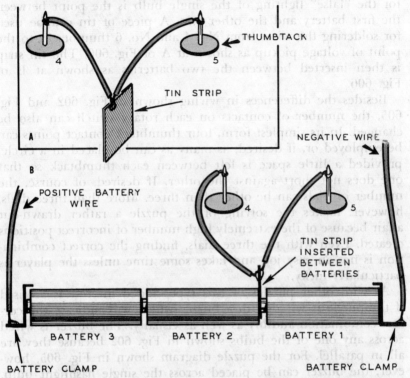

Fig. 606. *Method for tapping off a voltage from three batteries in series. The inset (A) at the upper left shows the detail while the remainder of the drawing (B) shows the entire battery arrangement.*

No. 3. In this particular puzzle one of the flashlight bulbs will light if dial No. 3 is rotated to No. 1, 4, 6 or 8 position. This will add a little more interest to the game because some bulb lighting will occur as the dials are rotated while trying to find the proper combination. The lighting of the one bulb for some positions of dial No. 3 does not, of course, give any clue regarding the proper combination but, since the player is not aware of this, he might interpret the bulb lighting as being of some significance. Because these positions are incorrect with respect to the proper solution, the solving of the puzzle is prolonged to a considerable extent.

For one flashlight bulb to light while the other two do not, it is necessary to tap off from the first battery as shown in Fig. 605. One bulb requires only one-third of the voltage furnished by the three batteries in series. If the voltage were tapped from the 4.5-volt source, the single bulb would burn out. Hence, the source for the "false" lighting of the single bulb is the point between the first battery and the other two. A piece of tin can be used for soldering the wires from No. 4 and No. 6 thumbtacks to the point of voltage pickup as shown at A of Fig. 606. The tin strip is then inserted between the two batteries as shown at B of Fig. 606.

Besides the differences in wiring shown in Fig. 602 and Fig. 605, the number of contacts on each rotary switch can also be changed. In its simplest form, four thumbtack contact points can be employed or, if desired, as many as can be placed in a circle, provided a little space is left between each thumbtack so that one does not short against the other. If desired, of course, the number of dials can be other than three. More than three dials, however, makes the solving of the puzzle a rather drawn-out affair because of the extremely high number of incorrect positions created. Even with the three dials, finding the correct combination is not an easy job and takes some time unless the player is particularly lucky.

As with other puzzles of this type, added interest is aroused if the puzzle is wired with a buzzer which will indicate the correct combination audibly as well as visually. The buzzer is wired across any one of the bulbs shown in Fig. 602 because they are all in parallel. For the puzzle diagram shown in Fig. 605, however, the buzzer can be placed across the single flashlight bulb. The buzzer will sound every time this bulb lights, both for false indications as well as when all bulbs light up. For details on buzzers, see Chapter 8.

Adding machine puzzle

The slide-contact rotary type of switch described for the previous puzzles can also be employed for an Adding Machine puzzle shown in Fig. 607. While this bears some similarity to the Mystery Safe game just described, it is entirely different with respect to finding the combination. Here the object is to find a combination of numbers which adds up to 12. There are, of course, many number combinations of the three dials which will give the sum of 12, such as 6 — 3 — 3, 5 — 4 — 3, etc., but only the proper combination will light the indicator bulb. Because of the necessity for adding the numbers shown on the dials, this puzzle will be of some benefit to youngsters, besides providing entertainment, because it will give them practice in addition. This does not mean that the puzzle will interest only youngsters. Actually the challenge of finding the right combination of numbers to solve the puzzle makes it intriguing to all age groups.

In contrast to the previous puzzles, the numbers on the dials are not arranged in orderly fashion but are mixed for the sake

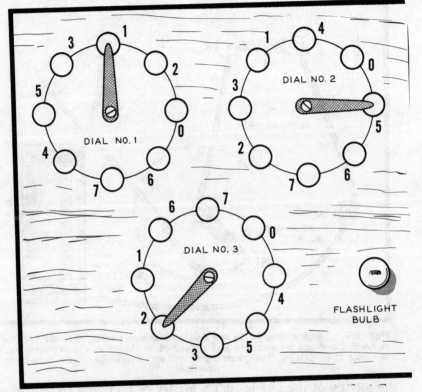

Fig. 607. *Front view of the Adding Machine puzzle. It uses the same type of rotary switch as in the Mystery Safe puzzle. The numbers near the dials are deliberately mixed to add interest in solving the puzzle.*

of variety. Also, a zero has been included so that the player might suspect that a combination such as 6 — 6 — 0 might be the solution.

The wiring diagram for the combination 4 — 3 — 5 is shown in Fig. 608. For simplicity, the thumbtacks not used are not shown. Note that the proper combination must be No. 4 tack on dial No. 1, No. 3 on dial No. 2, and No. 5 on dial No. 3.

If the dials are set on 3, 5, 4 for dials 1, 2, and 3, the indicator bulb will not light even though the sum of the numbers is also 12. The right numbers on the right dials must be found to light the bulb. Also, for other combinations which give 12, such as 6 − 6 − 0 or 2 − 3 − 7, the indicator will not light. Hence,

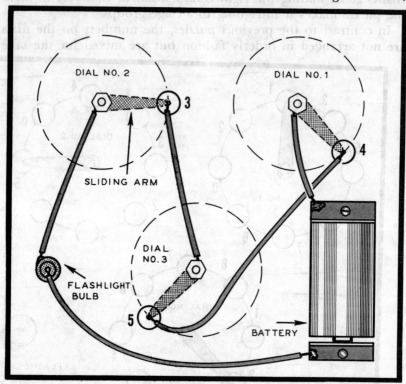

Fig. 608. *Wiring diagram of the Adding Machine puzzle. A single battery is needed and can be mounted and connected as shown in the drawing. For the sake of simplicity, those thumbtacks not wired into the circuit have been omitted.*

the solving of the puzzle isn't a matter of finding a combination of numbers which adds up to 12, but rather the *right sequence* of numbers.

Thumbtacks, sliding arms, etc. are assembled according to instructions for the previous puzzles. A single flashlight bulb is shown in Fig. 608, though two or more in parallel can be used for greater visual effect. Also, the inclusion of a buzzer across the flashlight bulb will be of added interest when the puzzle is solved. (See Chapter 8.) Once the combination is known to the players it can be changed by rewiring or by use of the double-pole double-throw switch type of change illustrated for the Mystery Safe puzzle in Fig. 605. Also, as with that puzzle, the number of thumbtacks used for each dial can be other than the eight shown in Fig. 607. The more numbers for each dial, of course, the more difficult it is to solve the puzzle. It is better to have only about eight for each dial so as not to make the solving of the puzzle too long or too difficult. Even with eight tacks for each dial, it often takes some time before the right combination is found.

Magic number puzzle

A puzzle particularly interesting to the player is one using a special switch which is tripped by use of a magnet. The switch

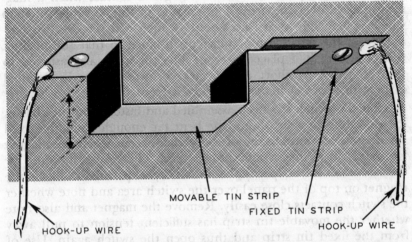

MOVABLE TIN STRIP

FIXED TIN STRIP

HOOK-UP WIRE

HOOK-UP WIRE

Fig. 609. *Details of the magnetic switch. The movable tin strip should be as close as possible to the fixed tin strip, but without touching it. Placing the magnet above the fixed tin strip will close the switch.*

itself is not visible from the top panel and to close the switch a small magnet must be placed directly over the area of the switch location. The switch itself is easy to construct. The only requirement is that several small bar magnets or magnet squares be available. Small magnets of this type can be procured from electronic wholesale houses or may be salvaged from discarded PM (permanent-magnet) type radio loudspeakers. Old speaker magnets are particularly preferred. The magnets can also be procured from some toys.

When constructing a puzzle utilizing the magnetic principle for the switch, the top panel *should not be more* than 1/4 inch thick so that when a magnet is placed over the magnetic area it will have sufficient pull to trip the switch. The switch itself is formed from tin strips ½ inch wide. One tin strip is bent in the form of a rectangle as shown in Fig. 609 to provide a certain degree of flexibility while at the same time exposing only a small area to the magnet. If a flat tin strip were used, the magnetic area would be too large. As shown in Fig. 609, the movable tin strip forms one switch contact. When a magnet is placed over the magnetic area, the movable tin strip will be pulled upward and its end will make contact with a fixed tin strip as shown. The end of the movable strip should not overlap the fixed tin strip by more than ⅛ inch. If approximately ¾ inch remains in the flat end of the movable tin strip, sufficient surface for magnetization will be available for closing the contact when a magnet is placed over the area. The portion of the movable tin strip which is bent downward for about ½ inch will remove this portion from the panel top a sufficient distance so that the magnet will not affect it if placed over it. The tin strips are fastened to the top panel by short carpet tacks or woodscrews as detailed earlier for other switches constructed from tin.

After the switch has been assembled and fastened to the chassis, the movable tin strip should be bent far enough from the fixed tin strip to prevent closing of the switch except when the magnet is directly over the switch contact area. The proper tension for the movable tin strip can be found by experiment. Place the magnet on top of the panel over the switch area and note whether the switch contacts close easily. Remove the magnet and also note whether the movable tin strip has sufficient tension to pull away from the fixed tin strip and thus open the switch again. Use of a magnet in the vicinity of tin may cause the tin to become slightly magnetized. If the movable tin strip does not have sufficient spring tension it will stick to the fixed tin strip even after the magnet has been removed. When this occurs, of course, the switch fails to open.

One application of the magnetic type switch is the construction of the Magic Number puzzle shown in Fig. 610. Here numbers are placed in random order in the squares marked off on the top panel. A single indicator bulb shows when the puzzle has been solved. The player is supplied with three magnets and the object of the game is to place each magnet in one of the squares of the puzzle so that the numbers in the three occupied squares total 13. As with the Adding Machine puzzle previously described, a number of combinations will give the correct sum, hence it takes some time to arrange the magnets in certain squares to solve the puzzle.

The wiring for the Magic Number puzzle is simple; only three switches need be constructed and hooked up. Fig. 611 in-

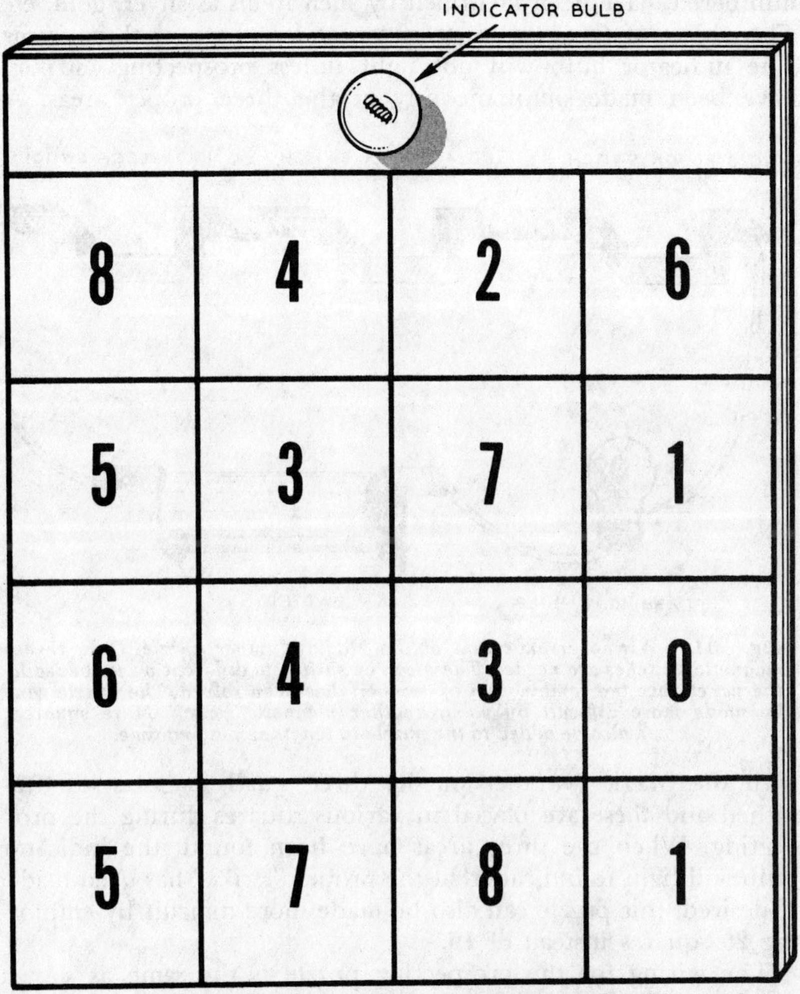

Fig. 610. *The Magic Number puzzle uses magnetic switches. The numbers on the puzzle are arranged at random.*

dicates the hookup for these switches, the flashlight bulb and the battery. The switches are fastened beneath the chosen number areas and then wired to the bulb and battery as shown. The choice of squares is left to the constructor and either of the No. 8 blocks can be used, as well as either of the No. 4 or 1 blocks.

Prospecting puzzle

The Magic Number puzzle, like others described in this chapter, lends itself readily to variations in basic design. One such modification is shown in Fig. 612. The squares, instead of being numbered, have been identified by such items as silver, gold, etc. The object of the game is to prospect for three of these items. The indicator bulb will not light unless prospecting "strikes" have been made simultaneously in the three proper areas. As

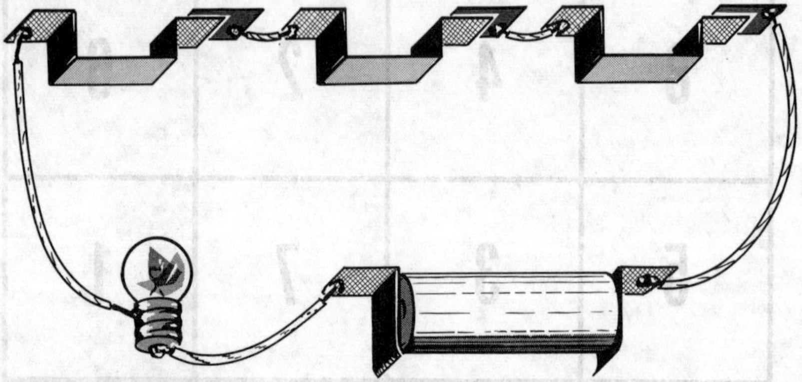

NO. 8 BLOCK SWITCH   NO. 4 BLOCK SWITCH   NO. 1 BLOCK SWITCH

FLASHLIGHT BULB   BATTERY

Fig. 611. *Wiring arrangement of the Magic Number puzzle. Only three magnetic switches are needed. These can be shifted to different places beneath the panel once the combination of numbers has been solved. The puzzle can be made more difficult by using another magnetic switch. More squares can also be added to the puzzle to lengthen playing time.*

with the Magic Number puzzle, three small magnets are furnished and these are placed in various squares during the prospecting. When the three areas have been found, the indicator bulb will light to indicate that the proper "strike" has been made. If desired, this puzzle can also be made more difficult by employing 25 squares instead of 16.

The wiring for the prospecting puzzle is the same as shown for the Magic Number puzzle in Fig. 611. The switches are installed beneath the three areas chosen by the constructor and may be one each of the silver, gold and iron squares. If the puzzle is to be built for youngsters only, it can be made more attractive by using pictures of apples, oranges, etc. instead of words in the squares. The purpose of the game would be to select the three fruits which will cause the indicator bulb to light.

Hidden word puzzle

A more elaborate version of the Magnetic Switch puzzle is shown in Fig. 613. Here, a total of 25 squares is used, with various letters placed in the squares as shown. The player is assigned four magnets instead of three as in the previous two puzzles. The

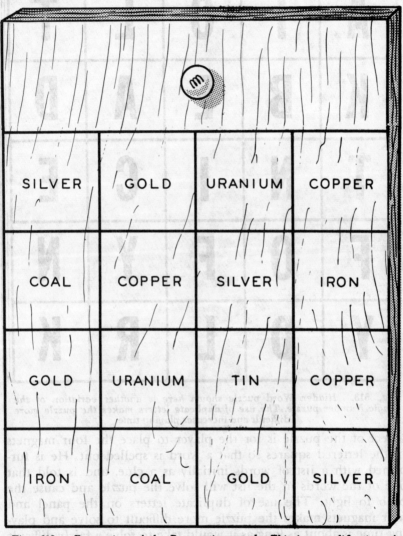

Fig. 612. *Front view of the Prospecting puzzle. This is a modification of the Magic Number puzzle. Although only 16 metals are shown here, the puzzle can be expanded to a total of 25. This type of puzzle lends itself very readily to numerous changes. Pictures can be used instead of words, or combinations of pictures and words.*

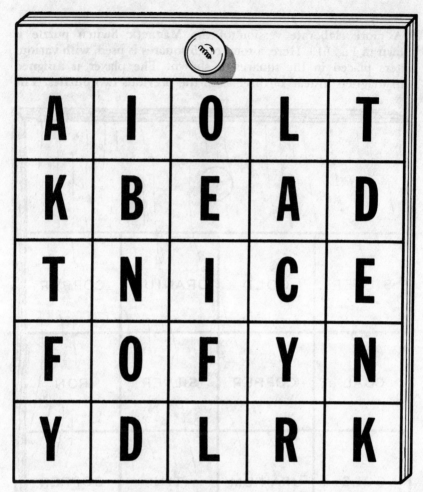

A	I	O	L	T
K	B	E	A	D
T	N	I	C	E
F	O	F	Y	N
Y	D	L	R	K

Fig. 613.   *Hidden Word puzzle shown here is another variation of the Magic Number puzzle. The use of duplicate letters makes the puzzle more difficult and increases playing time.*

object of the puzzle is for the player to place the four magnets in the lettered squares so that a word is spelled out. He is furnished with a list of words initially as a clue, and is told that one of the words in the list will solve the puzzle and cause the bulb to light. The use of duplicate letters on the panel and four magnets makes the puzzle more difficult to solve and playing time is about as long as it would take to solve a fairly difficult crossword puzzle. For the group of letters shown in Fig. 613, the following list of words can be employed as clues: read, book, beak, look, lock, leak, deer, cook, tiny, find.

Many more words could be given as clues, but the more words given the more difficult it will be to solve the puzzle. Initially, all the words listed can be given at the beginning of the game. If the player has difficulty in solving the puzzle within a reasonable time, to facilitate the solving of the puzzle, he can be told that the word is among the first five, with "look" the key word which will solve the puzzle, provided the proper letters are chosen from those given. Since there are duplicate letters on the panel, it will still take a little time to solve the puzzle.

The wiring for this puzzle is shown in Fig. 614 and follows the same pattern given in Fig. 611 for the Magnetic Number

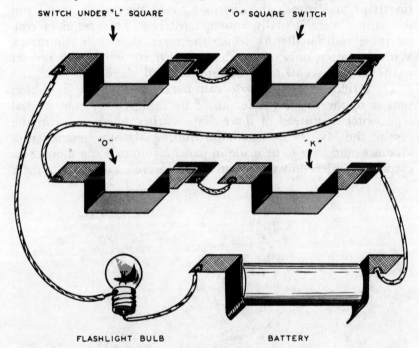

Fig. 614. *Wiring diagram of the Hidden Word puzzle. The magnetic switches can be shifted to different locations when the puzzle has been solved.*

puzzle, except that an additional switch is wired in series with the three shown in Fig. 611. The four switches are mounted under the letter areas chosen to spell out the word.

The alphabet feature of this puzzle makes the game particularly attractive for youngsters because they are given clues in terms of a word list beforehand. Hence, this Hidden Word puzzle will have an educational value for youngsters in a manner similar to the numeric type of puzzle such as the Magic Number and the Adding Machine puzzles. Because the puzzle is not too easy to solve, however, it is also of considerable interest to those in the older age group.

The Magic Number, Prospecting and Hidden Word puzzles could, of course, be constructed using plunger type switches, but this is not advised because the player can look into the holes of the plunger switches and see the contacts which will make the electrical connections. Thus he could solve the puzzle without trying the switches. To eliminate this possibility when using plunger switches, all the holes would have to be installed with tin strips underneath the chassis so that the player could not tell which were actually contact switches. This requires considerable additional work which the magnetic switch eliminates. With the latter, only those squares which are active with respect to the puzzle have switches mounted beneath them.

The Hidden Word puzzle can be constructed in a fashion similar to the Mystery Safe puzzle by using rotary contact dial type switches. Instead of three dial switches, however, as in the case of the Mystery Safe and the Adding Machine puzzles, provisions would have to be made in panel layout to accommodate an extra dial, unless only three letter words were used for the puzzle.

miscellaneous games

T HE games that are described in this chapter illustrate the manner in which variations can be made with respect to the basic unit. This will enable the reader to undertake some original design work. Miscellaneous games of a basic nature are described and possible changes from original design are given to indicate how to increase playing interest, etc.

Satellite game

The Satellite game illustrated in Fig. 701 lends itself readily to changes in construction. As shown the game is designed for four players although it could also be constructed for as many as six while still keeping the game a reasonable size. Since this game can be built in a manner chosen by the reader, no layout dimensions are given. Depending on the size of the metal plunger type switches employed and the spacing between them, the top panel can be formed from a section 8 inches square to 12 inches square or more.

Five circles are drawn on the top panel and around the circumference of each circle are four holes. The holes of the four inner circles are for the metal plunger type switches. The holes in the outer circle are for mounting flashlight sockets.

Each player is allocated four metal plungers for the plunger switch holes, and one flashlight bulb. The object of the game is for each player to try to launch a satellite in its orbit before the other players do. The game starts after deciding which player goes first. The first player spins the pointer shown in Fig. 702.

He is not allowed to place any plunger (or bulb) in the holes until he spins to the "start" position on the spinner dial. If the pointer (or spinner) stops at any other indication, the player stops and the next player has his turn, etc.

Once a player has landed on the start section of the spinner dial, he places one of the plungers in the first inner ring, in the hole corresponding to his chosen number. With a plunger on the earth circle, he may proceed with the rest of the game as his

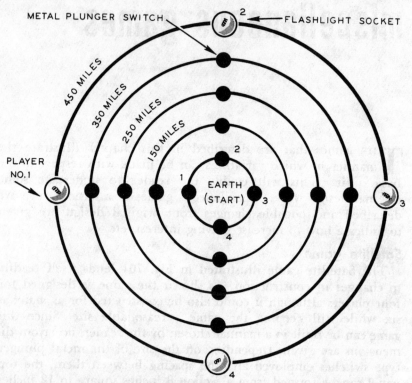

Fig. 701. *Suggested layout for the Satellite game. This game uses metal plunger switches of the type described in an earlier chapter.*

turn comes up on the dial. After a player has officially started by having one plunger in the earth circle, he may insert another plunger (or the flashlight bulb) in place as indicated by the spinner. If, for instance, the pointer lands on the 250-mile indication, he inserts one of the metal plungers into the hole in the 250-mile orbit of the game. If the spinner stops on the 450-mile indication of the spinner dial, he screws the flashlight into its proper socket. The flashlight, however, will not light until all plungers are in place. Also, if the spinner lands on the start section or on another section where the player already has a plunger, he does nothing and simply misses a turn. The first player with all the plungers in the various orbits, plus the flashlight socket, will have launched his satellite and the bulb which represents the satellite will light. The other players can continue,

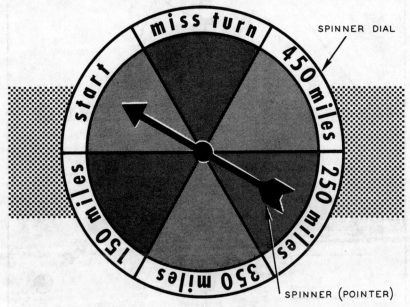

Fig. 702. *Spinner dial and pointer for the Satellite game. The dial is similar to those used in the other games in this book. Mount the spinner so that it rotates freely.*

if desired, to see who is second, third, etc. (The plungers can be indicated as the various stages of the launching rocket, such as the first stage, second stage and third stage.)

The wiring for the Satellite game is shown in Fig. 703. The vertical and horizontal wiring sections in solid lines represent the basic game. If additional lanes for other players are to be incorporated in the game, they should be wired as shown by the diagonal dotted section. Metal plunger switches as used in previous games are employed, and a single battery is sufficient for this game.

The flashlight bulbs are all in parallel with the battery and, when all are in place, the current drain from the battery will be high. Consequently the bulbs will not light as brightly. Since, however, the bulbs are on only during the latter part of the

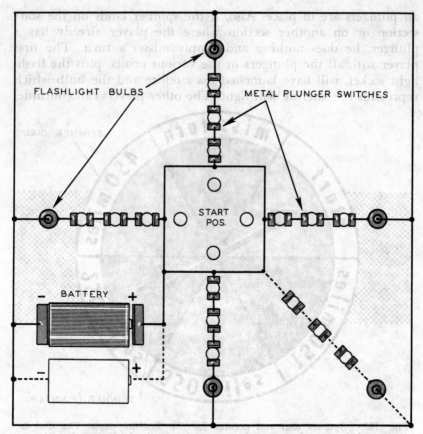

FLASHLIGHT BULBS

METAL PLUNGER SWITCHES

START POS.

− BATTERY +

− +

Fig. 703. *Wiring arrangement of the Satellite game. Two or more batteries can be connected in parallel as shown by the dotted section in the lower-left corner of the drawing.*

game, there is no prolonged drain on the flashlight battery. For longer life, two or three flashlight batteries can be used in parallel; that is, additional batteries are connected to the first one with all plus terminals connected together and all minus terminals connected together. This is shown by the dotted section in Fig. 703.

Playing instructions—satellite game

1. Initially, each player is assigned four metal plunger pegs and one flashlight bulb. He is also assigned a number to indicate the path of his play on the board.

2. The first player spins the pointer to indicate the move to be made. If the dial indicates start, he places a plunger in the hole in the inner circle marked Earth — Start, in the

hole corresponding to his number. If the spinner dial lands on any other section, he does nothing and again awaits his turn.

3. Other players take turns in order and do not place a plunger in any circle until they spin the pointer to the start position.

4. After a player has a plunger in the start circle, he places other plungers in the respective orbits as indicated each time his turn comes up.

5. The first player who has all plungers (as well as his flashlight bulb) in place wins the game as indicated by the lighting of his satellite.

6. Other players continue to see who launches the second satellite, the third, etc.

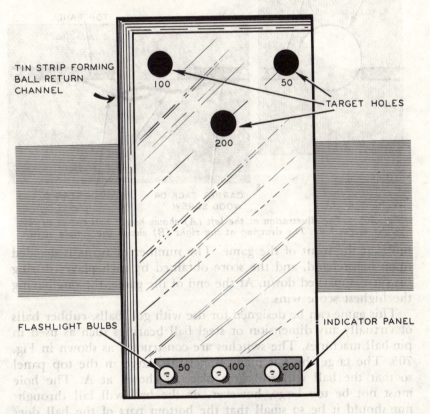

Fig. 704. *Physical layout of the Tri-target game. This is a top view, looking down on the game. When a ball lands in the target hole it closes a switch. This will turn on one of the flashlight bulbs in the indicator panel at the lower front end of the game. The size of the target hole will be determined by the size of the balls that are used.*

Tri-target game

Another game which lends itself readily to the design ideas of the builder is the Tri-target game illustrated in Fig. 704. It is based on the pin-ball idea. Balls are rolled down the top panel and the object is to have a ball come to rest in one of the target areas. Three target areas are shown, although the game could be built with only a single target or with four or five if desired.

Each player is assigned three balls which he uses during his turn. The object is to roll the balls down the panel and have them come to rest in one of the targets. When a ball stops in a target area, it will trip a switch, lighting one of the scoring

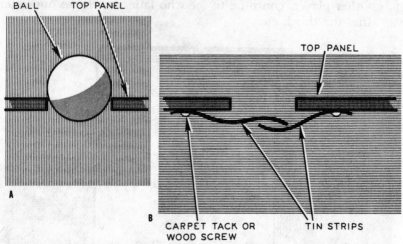

BALL   TOP PANEL

TOP PANEL

A

B

CARPET TACK OR   TIN STRIPS
WOOD SCREW

Fig. 705.   *The illustration at the left (A) shows how a ball comes to rest in the target hole. The drawing at the right (B) shows the simple switch.*

bulbs at the front of the game. The number of turns is decided upon beforehand, and the score obtained by each player during his turn is marked down. At the end of the game, the one having the highest score wins.

This game can be designed for use with golf balls, rubber balls of virtually any dimension or steel ball bearings such as used in pin-ball machines. The switches are constructed as shown in Fig. 705. The target hole is made sufficiently large in the top panel so that the ball will roll into place as shown at A. The hole must not be too large, however, or the ball will fall through; nor should it be so small that the bottom part of the ball does not project through the bottom panel. The switch is a simple type constructed as shown at B of Fig. 705. Two tin strips are cut and bent into shape as shown so that they do not make contact until a ball rolls into position. The weight of the ball will then apply sufficient pressure to close the switch contact. The width and length of the tin strips will also depend on the hole size and the weight of the ball. A little experimenting will quickly establish the relative size of the tin strips so that they have sufficient spring action to open when the ball is removed. A slight bending of the tin strips will adjust the tension so that they will function properly.

The panel layout and dimensions will also depend on the size of the balls used. With balls as large as those used in golf, a panel

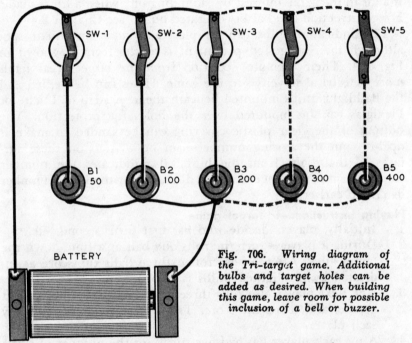

Fig. 706. *Wiring diagram of the Tri-target game. Additional bulbs and target holes can be added as desired. When building this game, leave room for possible inclusion of a bell or buzzer.*

section can be constructed measuring approximately 8 by 24 inches. A channel formed of tin (or ¼-inch plywood) is built along the far end and one side of the chassis to return those balls which do not go into the target holes. Side supports should be constructed so that a small chassis is formed to permit housing the battery, the switches and the flashlight-bulb sockets beneath the top panel. Only three balls need be on hand since each player uses the same ones when his turn comes, simply picking up those on the return tray or lifting them from the target areas.

The wiring for the game is shown in Fig. 706. The target area switches are designated as SW-1, SW-2, and SW-3, while the indicator bulbs are B1, B2, and B3. The particular placement of the target areas is left to the choice of the builder although the layout shown in Fig. 704 can be employed. Usually it is more difficult to get a ball in the center target, hence this area should be allocated 200 points. The areas toward the side of the top panel can be designated as 50 and 100 points, or both could be 50 points if desired. If additional target areas and scoring bulbs are to be included, they can be wired in parallel with the others as shown in the dotted sections of Fig. 706. As with other games and puzzles described earlier, if a buzzer is wired across each bulb it adds to the game interest by "sounding off" when a hit is made. For construction of a battery-operated buzzer see Chapter 8.

The indicator bulbs can be placed in any convenient spot although their suggested placement is at the front as shown in Fig. 704. Their "remote" position from the target areas lends more "electrical novelty" to the game. Holes can be drilled and the flashlight bulbs mounted beneath them. A strip of Lucite or Plexiglas can be mounted over the holes for protection. The bottom of the clear plastic covering can be sanded to make it opaque and the scoring numbers can be written on these surfaces so that light from the bulbs also indicates the number scored. A similar arrangement was discussed for the Little Thinker described earlier.

Playing instructions—tri-target game
1. Initially, players decide who has first turn, second, etc.
2. During a player's turn, he rolls one ball at a time down the target alley. After he has rolled three balls the score as indicated by the lighted bulbs is recorded.
3. The next player also rolls three balls down the alley, one at a time, and totals the score. This procedure is repeated by each player.
4. After each player has had five turns (or the number of turns decided upon beforehand) the game ends and the one having the highest score wins.

Basket-catch game
The type of switch described for the Tri-target game can be adapted readily to include many target areas and more scoring lights to resemble the typical pin-ball game.

Another application of such a switch type is the Basket-Catch game illustrated in Fig. 707. Here, each player is assigned several rubber balls. Each player, in turn, stands a prescribed distance from the basket and tosses the balls into the basket. Inside the basket is a chassis, such as shown in Fig. 708, which has target hole switches. The player getting the highest score in a predetermined number of tosses, wins.

For the Basket-Catch game, a wooden or plastic wastebasket is used. A game chassis is then built which will fit inside the basket so that it rests on the bottom. The game chassis has four target holes with switches attached beneath them on the underside of the chassis, as illustrated for the Tri-Target game in Fig. 705. The indicator flashlight bulbs, however, are mounted on

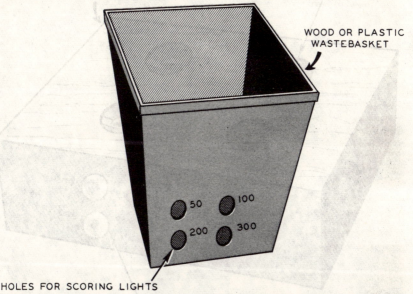

Fig. 707. *An ordinary wastebasket can be used when building the Basket-Catch game. Use any non-metallic type of basket.*

the chassis side as shown in Fig. 708. The chassis is wired for four switches as shown in Fig. 706, ranging from 50 points to 300. Holes are then cut into the wastebasket so that they line up with the flashlight indicator bulbs on the chassis. The chassis is then placed in the wastebasket and the device is ready for playing.

Small rubber balls are preferred so that the target areas on the chassis top need not be too large. If the target area holes have to be enlarged too much, scoring will be too easy on a small chassis. If a fairly large-sized wastebasket is used, the rubber balls can be somewhat larger. If desired, only two or three target areas can be used instead of four. Fewer target areas are preferred if the wastebasket area at the bottom of the basket is not too large.

Three balls can be assigned for the playing of the game, each player retrieving the balls after he has tossed them into the basket. The game scoring limit could be set at 1,000 points and the first player to get this number wins. Because the same three balls are used by each player, there is no limit to the number of players who can participate.

The playing instructions which follow for the Basket-Catch game are suitable for a game having four target areas and using

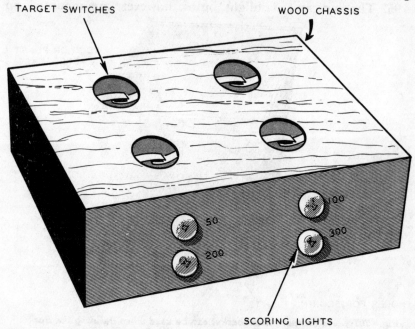

Fig. 708. *Chassis for use in connection with the Basket-Catch game. Although only four target areas are shown, more target switches can be added. The pressure of the ball closes the switch.*

three balls. The rules can, of course, be changed to suit the particular game constructed.

Playing instructions—basket-catch game
1. Initially, the players decide who has first turn, second, etc.
2. Each player, during his turn, must stand 8 feet from the basket.
3. Each player, during his turn, tosses one ball at a time into the basket, until he has tossed three balls.
4. (Optional) Any player scoring 600 during his turn, is permitted to toss an extra ball for additional scoring.
5. Any player who first reaches a score of 1,000 wins the game.

The magnetic switch described in Chapter 6 and used with the Magic Number puzzle and others described there (illustrated in Fig. 609) is also adaptable for games. Construction is simpler in a game using magnetic switches because there is no necessity for drilling all the holes required for metal or wood plunger type switches. As described in Chapter 6, however, small magnets must be obtained. The magnets should be approximately $\frac{1}{2}$ inch square or $\frac{1}{2}$ inch in diameter so that the panel dimensions of the game can be kept to a minimum. If, however, only larger

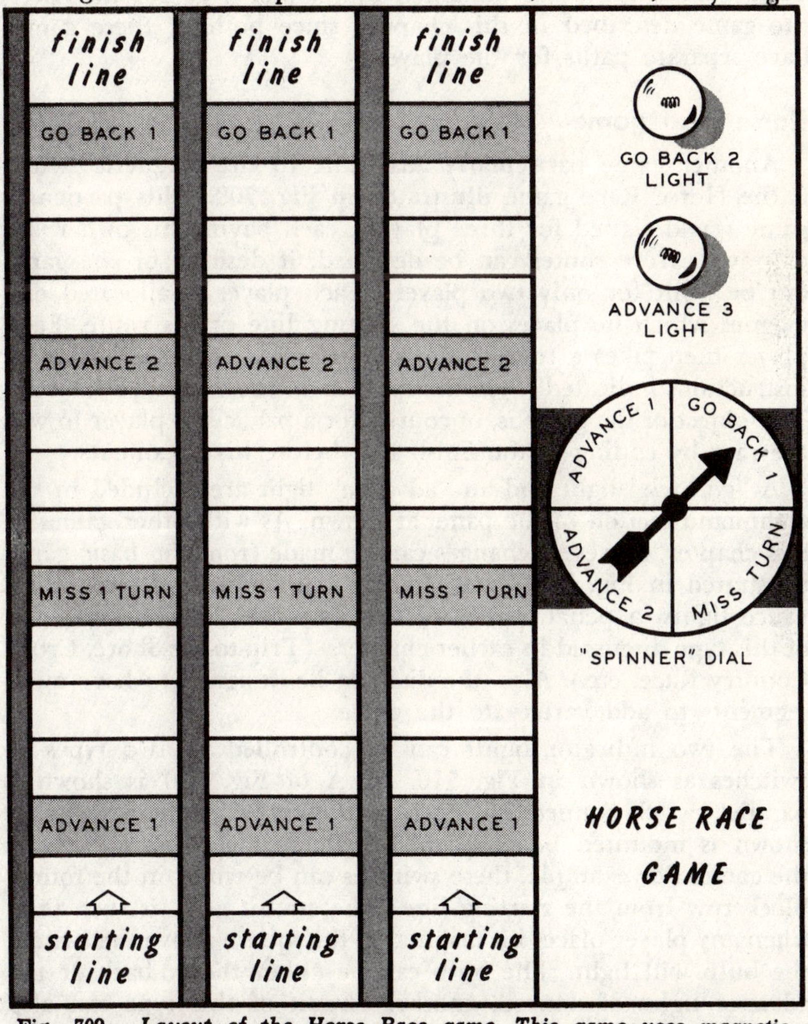

Fig. 709. *Layout of the Horse Race game. This game uses magnetic switches described in the preceding chapter. The use of a magnetic switch eliminates the need for drilling holes.*

type magnets (1 inch in diameter) are available, the game panel will have to be made larger so that there will be no interaction between the magnets of various players. To prevent interaction and magnet clinging, any game constructed with magnet switches should be such that each player has a route or path of his own. If all players use the same path, it will be virtually impossible to place magnets near each other because they will pull together.

The magnetic switch can be employed for such games as the Trip-to-the-Shore game described in Chapter 4 or for the Satellite game described in this chapter, since both of these games have separate paths for the players.

Horse race game

Another game particularly adaptable to the magnetic switch is the Horse Race game illustrated in Fig. 709. This particular game was designed for three players, each having his own route of travel. More routes can be designed, if desired, or the game can be built for only two players. Each player is allocated one magnet which he places on the starting line of his route. Each player then takes a turn at the spinner and moves according to instructions indicated when the spinner on the dial stops rotating. The object of the game is, of course, for a particular player to win the race by ending at the finish line before his opponents.

A "go-back" light and an "advance" light are included in the right-hand section of the panel as shown. As with other games in this chapter, however, changes can be made from the basic game illustrated in Fig. 709. Such changes can include additional advance lights, a penalty area, etc. as employed in similar games of this type discussed in earlier chapters (Trip-to-the-Shore, Cross-Country Race, etc.). Also, the dial can be designed to have more segments to add variety to the game.

The two indicator lights can be controlled by two types of switches as shown in Fig. 710. At A of Fig. 710 is shown a parallel switch connection. Each of the three magnetic switches shown is mounted beneath one of the segments or squares of the game. For example, these switches can be wired in the fourth block row from the starting line. The circuit at A is such that, when any player places his magnet on the square above the switch, the bulb will light. The bulb can be either the go-back or the advance light. If desired, another circuit of this type can also be placed in the ninth row from the starting line.

A series circuit is in B of Fig. 710. Here, the bulb will not light unless all players occupy the same row of spaces. This type of switching adds some variety to the game because, if one or two players land on this position, the bulb will not light. When the third player lands on the same position, however, the bulb will light either to indicate an "advance" or a "go-back" as the constructor of the game desires. Again, this switching arrangement can be added to several rows, such as the first row after the starting line and perhaps two other rows somewhere along the

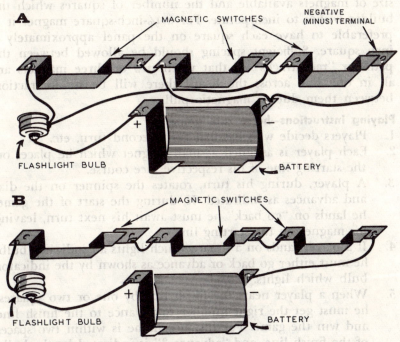

Fig. 710. *The drawing at the top (A) shows magnetic switches in parallel. The lower drawing (B) illustrates magnetic switches in series.*

route. When several of the switch arrangements shown at B are employed, each game is somewhat different from the preceding one because the player landing over one of the magnetic switch areas may or may not receive a penalty, depending on the position of the other players. Regardless of how many switches are used, a single battery is employed. To prevent errors in wiring it is preferable to connect all the negative leads to the negative side of the single battery, as shown in A of Fig. 710.

The panel can be fastened to sides 1 to 2 inches high so that the battery and switches can be mounted conveniently underneath the panel. The squares of the game should be painted various colors to make for an attractive appearance. The indicating lights can be capped by a colored plastic dome or jewel as described for earlier games. Construction of the spinner dial is also covered in earlier games.

For effective functioning of the magnetic switches the top panel should preferably be of ¼-inch plywood. The exact dimensions of the top panel are not given because they will depend on the size of magnets available and the number of squares which the builder wishes to incorporate. For a ½-inch-square magnet it is preferable to have each square on the panel approximately 1 inch square. Sufficient spacing should be allowed between the paths or "race tracks" so that when two or three magnets are all in one line across the board there will be no interaction between them due to magnetic pull.

Playing instructions—horse race game

1. Players decide who has first turn, second turn, etc.
2. Each player is assigned a small magnet which he places on the starting line of his respective race course.
3. A player, during his turn, rotates the spinner on the dial and advances as indicated. If, during the start of the game, he lands on "go back" he must await his next turn, leaving his magnet on the starting line.
4. If a player lands on a square which lights an indicator bulb, he must either go back or advance as shown by the indicator bulb which lights.
5. When a player nears the finish line by one or two squares, he must get the right number to advance to the finish line and win the game. For instance, if he is within two spaces of the finish line, and "advance 3" is indicated by the bulb, he must wait until he gets an "advance 2" on the spinner dial.

puzzle and game accessories

IN all of the games described so far in this book, the components required have been held to a minimum—mostly switches, lights and batteries. Once these games have been constructed, however, their general appeal can be increased to a considerable extent by adding other accessories. The thinking type games, such as the Little Thinker and Twenty-one, can be improved by the addition of a delay so that, when the button is pushed, the machine does not give the answer immediately but seems to require a little time to "think" out the proper number of pegs it wants removed. Similarly, such games as the Trip-to-the-Shore, Jet Plane and others, have greater interest if a buzzer is made to sound simultaneously with the lighting of the indicator bulbs. Hence, while all these games and puzzles are complete in themselves, each can be improved by incorporating one or more accessory items.

Dc buzzers

Since all the games described operate on flashlight batteries, a buzzer arrangement calls for the use of a unit which is sensitive to the low voltage employed while still providing a sufficiently loud buzz to be heard clearly. A simple buzzer of this type can be constructed from strips cut from ordinary tin cans and from a coil wound with easily obtainable cotton- or enamel-covered wire.

The buzzer should be constructed on a separate piece of wood as a panel support and tested for operation before installing and wiring into the complete games and puzzles. A section of ½-inch plywood measuring 4 inches by 2 inches has been found adequate for mounting the buzzer. The coil requires an iron core so that it will become a stronger electromagnet when in operation. The core for the coil consists of a section of a common nail. A 2½-inch nail (8-penny nail) is used, though virtually any nail will work equally well. Several types of buzzers have been constructed by the writer and all have worked well even though nails of various sizes were employed. For the 2½-inch nail, cut off the pointed

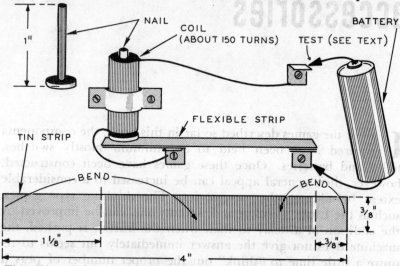

Fig. 801. *A buzzer can be made by winding about 150 turns of wire around a nail. When a current passes through these turns of wire, the coil becomes an electromagnet. The flexible strip is fastened at one end only and is free to move back and forth at the other end. The movement of the flexible strip makes and breaks the circuit.*

end so that the remaining section measures approximately 1-inch as shown in Fig. 801.

The coil is formed by using either No. 26 or No. 28 double-cotton-covered copper wire which can be obtained in a spool from radio parts stores or radio parts catalogs. Enameled or plastic-coated wire can also be used. The coil can also be wound with thinner wire and still function fairly well. Before winding the wire on the nail, wrap a single layer of Scotch tape around the nail as insulation so that the copper wire will not short against the nail in case the insulation is penetrated by any rough spots on the nail. Wrap approximately 150 turns of wire on the nail, leaving several inches of wire sticking out of the coil when starting to wind it. After the turns of wire have been placed on the coil, leave another 2- or 3-inch section of wire for a connecting lead and cut off the wire from the spool. Cover the coil thus formed with Scotch tape to prevent the wire from unraveling and to form an insulation over the entire coil.

Cut a piece of tin measuring approximately ⅜ inch wide by 1¾- inches long to act as a clamp for the coil. Bend this tin strip around the coil and punch holes in each end with an icepick or awl. Use ¼-inch wood screws or tacks to fasten the coil down at one end of the mounting board as shown in Fig. 802.

For the buzzer arm, a ⅜-inch-wide strip of tin is cut to a length of 4 inches. Bend over 1⅛ inches of the end on itself,

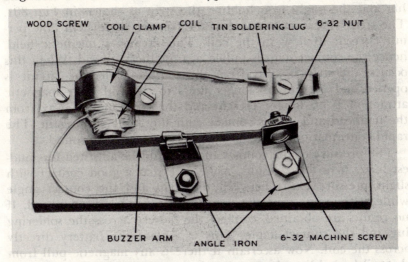

Fig. 802. *A completely assembled buzzer. The use of a buzzer adds excitement and interest, and can be added to any of the games described earlier. The wire wrapped around the nail can be kept in place with Duco cement or Scotch tape.*

pressing the crease tightly together with pliers. The edge of this folded strip should now be soldered so that this end of the strip becomes a rigid double-thick section. This is the section which will be facing the coil as shown in Fig. 802.

Bend over ⅜ inch of the other edge and again solder the edges to make a rigid section. Drill a small hole in this short end to accommodate a 6-32 machine screw. An angle iron is employed for mounting the tin strip, using washers to position the strip so that it does not rub against the wood base and also to position it so that its flat end is opposite the nailhead. Use a machine screw and nut for fastening the strip to the angle iron. The angle iron can be fastened to the base by using a wood screw or another machine screw-and-nut arrangement. Space the flat end of the strip about ⅛ inch away from the nailhead.

Mount a contact angle iron just beyond the center of the strip, toward the coil, so that it is approximately at the end of the 1⅛-inch foldover as shown in Fig. 802. Bolt or screw this angle iron to the mounting board, adjusting it so that it applies a slight pressure on the strip. The pressure should be sufficient to bring the end of the strip within about 1/16 or 1/32 inch from the nailhead. One wire from the coil is now soldered to this contact angle, and the other wire from the coil is soldered to a tin soldering lug. Now, temporarily connect a battery across the soldering lug and the angle iron holding the strip as shown in Fig. 801. The strip (because it is against the contact angle) will now permit current to flow in the coil. This creates a magnetic field, hence the contact strip is pulled against the nailhead. When this occurs, the strip is away from the contact angle, the circuit is opened and current no longer flows to the coil. The magnetic attraction is no longer present and the strip moves away from the nailhead and makes contact with the angle iron again. The rapid continuation of this process creates a buzzing sound.

The pressure of the contact angle should be adjusted for loudest buzz. If no buzz results, make sure that a good contact with slight pressure exists between the strip and the contact angle while at the same time the strip is not touching the nailhead. If no results are obtained, connect the battery across the soldering lug and the contact angle. This will place the battery directly across the coil. Now ascertain if there is any magnetic pull from the nail by holding the metal end of a screwdriver to it. If there is a decided magnetic pull (make sure the screwdriver is not magnetized before making this test), the coil functions properly and failure of the device to buzz is because of incorrect tension or buzz contacts between the long strip and the contact angle.

The contact angle should preferably be made of some metal which makes good electrical contact such as tin or tinned, nickel-plated or chrome-plated copper. Coin-silver contacts can also be soldered into the strip end to the contact angle and will give much better service because of the better electrical connection.

Once the buzzer is in operation, the contact angle as well as the pressure of the strip can be adjusted very slightly to vary the pitch and the intensity of the buzz. When a suitably loud buzz is obtained, the buzzer is ready to mount in the puzzle or game. For the Jet Plane game, for instance, the buzzer is mounted directly across the light bulbs shown in Fig. 407. For the River Crossing game, the buzzer is mounted across the single light-bulb indicator shown in Fig. 306. For the Cross-Country Race and Trip-to-the-Shore type games, the buzzer can be wired across the advance indicator bulb. If a buzzer is to be used for other indicator lights, a separate buzzer will have to be constructed for each light indicator. A single buzzer cannot be used as a common

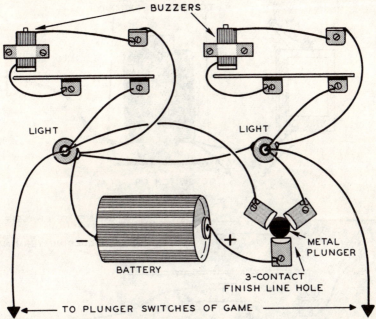

Fig. 803. *The diagram shows how a pair of buzzers can be operated from a single battery. It is best to add buzzers after a game has been completed and is working satisfactorily.*

buzzer for all the lights because it would interconnect the lights electrically and would result in all the lights turning on when the buzzer sounds.

The number of buzzers used depends on the choice of the constructor. For the Cross-Country Race, for instance, the writer used only two buzzers, one for the Advance-4 and one for the Back-3 light. However, when the finish line is reached and a plunger inserted into the last hole, a three-contact plunger switch is used so that both buzzers sound simultaneously to indicate completion of the game. The wiring for the three-contact switch as adapted to the Cross-Country Race game is shown in Fig. 803.

For the Trip-to-the-Shore game, a single buzzer was used across the Advance-5 light only, and another buzzer of a different pitch was used for the completion of the trip. If, however, buzzers are also desired for the back-up lights, individual buzzers are recommended. Use of individual buzzers, while involving a little more construction time, adds variety because each buzzer will have a

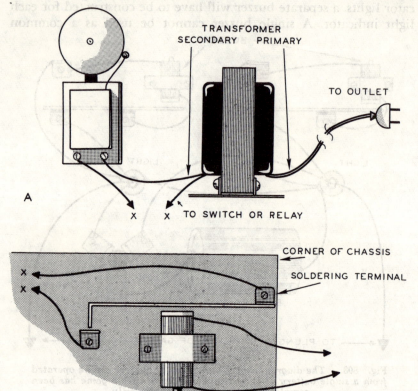

Fig. 804. *The upper drawing marked (A) shows how to connect a doorbell and transformer. The relay in the lower drawing (B) is a buzzer and is used for opening and closing the circuit to the doorbell.*

different pitch. The particular pitch of a buzzer will soon be recognized as indicating a certain type of move even before looking at the lights for identification of the move involved.

Ac buzzers

A much louder buzz than that obtained from the 1.5-volt buzzer can be obtained by using an ac type of buzzer or doorbell.

These higher-voltage buzzers, because they are so loud, will inject an element of surprise in the puzzle or game as well as indicating in no uncertain terms an error, etc. When such a loud buzzer is desired, it is necessary to wire the circuit with well-insulated wire. The most inexpensive buzzer can be made from a doorbell and an appropriate stepdown transformer. Both the doorbell and the transformer can be acquired in most of the larger dime stores and in virtually all hardware stores or in the hardware sections of department stores. The doorbell unit can be left intact to give a loud ring or the bell section can be cut off the doorbell with a

Fig. 805. *These are photos of commercial type 6-volt relays. They work in the same manner as the simpler kind of buzzer. Some relays will work from a single battery; others will require at least four batteries.*

hacksaw, thus forming a buzzer. The disadvantage with this type of indicator is that ac from the power lines of the home must be used, and some element of danger exists if the wiring is not fully insulated. Also, in most cases this arrangement is not practical to use directly in the circuits previously given unless a relay is employed. The system can, however, be used with the single-indicator type puzzles, such as the River Crossing or the Push-button Switch puzzles (double-indicator type).

The basic wiring diagram for the doorbell and transformer is shown at A in Fig. 804. The terminals marked X are for connection to the switch which is to ring the bell or buzzer, or from a relay as shown at B. The relay must have sufficient sensitivity to close with 1.5 volts. The input terminals of the relay shown are wired across the indicator light of any of the games discussed. When the voltage is applied to the light, it is also placed across the relay and closes the relay contacts. In turn, the switch closes the ac circuit of the bell transformer secondary and rings the doorbell or buzzer. The relay switch which closes the ac must be insulated from the input system (coil of relay) to prevent short circuits.

Small relays operating on battery voltages of from 1.5 to 6 can be acquired from hobby shops, radio parts stores, or some mail-order wholesale houses. If a 1.5-volt relay is not available, a larger one can be used. For instance, a 6-volt relay can be operated by placing four batteries in series. Typical 6-volt relays are shown in Fig. 605. These provide isolation between the relay tripping circuit and the controlled higher-voltage circuit. Relays should be purchased which will draw as little current as possible consistent with good relay action. The less operating current the relays require, the more economical they are with respect to battery replacement. If a higher-voltage relay is used, it will also be necessary to employ bulbs which will not burn out when using a higher voltage.

It is preferable to solder the connection between the line cord and the transformer. The soldered joints should then be covered with a good grade of electrical tape and then with friction tape.

The secondary of a bell transformer has low voltage, hence bell wire can be employed for wiring the circuit. If desired, however, radio hookup wire can be used and in most instances will have better insulation. This will minimize the danger of a short circuit occurring in the wiring. The use of a bell transformer and doorbell makes for an effective alarm system but does increase the cost of the puzzle. It also means that the puzzle must be plugged into an ac outlet. The ac type puzzle construction is advisable only when the games and puzzles are to be played by other than children. For the latter it is a much safer arrangement to use batteries instead of ac from the power lines.

Both the transformer and bell should be screwed or bolted to the sides of the cabinet so that each unit is rigid and will not move out of place. A wooden chassis acts as a tone chamber for the bell and provides a very loud signal.

Delay systems

Some of the puzzles described in this book can be improved considerably by the addition of a delay circuit to the indicating device. Two puzzles for which a delay system is particularly adaptable are the Little Thinker and Twenty-one. In these puzzles the "machine" is supposed to "think," hence a delay in the machine's choice of pegs to be removed makes it appear as though the machine must be given some time to think out its answer.

A delay system incorporated into either of these devices would work as follows: the player removes his choice of pegs and then depresses the button to indicate to the machine that he has made his selection. Instead of the lights going on immediately to indicate the machine's selection, there will now be a delay of approximately ¼ to ½ minute before the machine indicates its selection. Such a delay is quite effective. The personification of the machine's operation seems more valid because the machine now must take time to "think" what the proper selection is before it lights the flashlight bulbs.

Another puzzle in which a delay system can be incorporated is the River Crossing. Here, the delay system would not only pro-

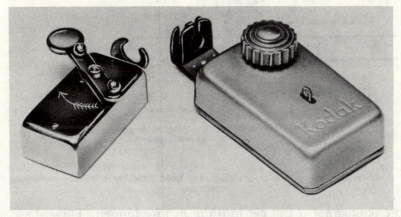

Fig. 806.   *Two types of automatic camera cable-release units. These can be used to provide a delaying action in the games. Units such as these can be purchased in most camera stores.*

long the operation of the puzzle but would heighten interest by making the player wait a few seconds after each move before he learns whether or not he has made an error. If, however, a delay system is incorporated in the River Crossing puzzle, a pushbutton would have to be incorporated in the design and the button depressed after each river crossing to ascertain whether the move is correct or incorrect. The delay system can also be applied to some of the other puzzles if desired.

Generally, however, the delay circuit does not particularly enhance the games. An exception is the Jet Plane game where a delay system could be used to slow down the action somewhat and introduce an element of suspense. An electronic delay system is difficult to design with electrical systems, using only a flashlight battery for the voltage source. A much more simple arrangement is a mechanical one where pushing the button winds a spring which, in turn, works a clock mechanism. The latter, in turn, will then trip a switch after a certain time interval. One method for doing this is to acquire an automatic camera cable-release unit available at camera stores. Two typical types are shown in Fig. 806. The one at the right is a standard Auto-release manufactured by the Kodak Co. while the one on the left is one of several foreign makes available at camera stores. These automatic release units sell for several dollars and are designed to trip a cable release on a camera automatically so that

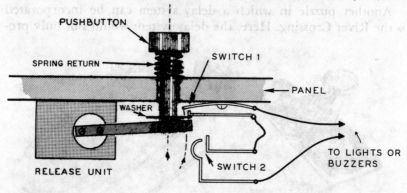

Fig. 807. *Method for using an automatic cable-release device. Two switches, as shown in the drawing, are used together with the automatic release.*

the photographer can get into the picture himself. These devices have a winding arm or knob which, when turned, winds up the internal spring. As the spring is wound, the trip lever at the top of the unit rises over an inch to permit the insertion of the cable release. A stop lever is provided so that the mechanism will not operate until desired. When the stop lever is tripped, the cable-release lever will then slowly return to its original position.

Such an automatic cable-release device is used as a delay system by mounting it beneath the pushbutton, on one of the side panels of the puzzle or game. When units have a lever arm, it is extended by soldering to it a 1½-inch-long piece of tubing or flat piece of metal. For those having a knob, an extension rod can be fastened to the knob by drilling two holes in the knob and tapping the holes to accommodate short 6/32 bolts as shown in Fig. 807. The knob on the Kodak unit can be screwed off counter-clockwise for convenience in drilling and tapping.

The pushbutton which is depressed to cause the puzzle or game to give an indication is so mounted that it presses down on the lever of the automatic release device. The stop lever of the mechanism remains in the on position. When the indicator lever is depressed, it will wind the spring mechanism and also cause the trip lever to emerge from the unit. The former is used for closing the switch. The switch itself is made from two strips of tin each measuring ¼ inch by 2 inches long, bent and mounted as shown in Fig. 807. The switch made from the tin strips is then wired across the indicator lights or buzzers of the puzzle.

The delay system must be further modified, however, because

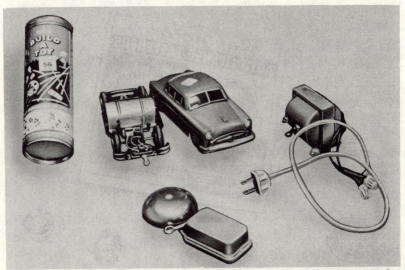

Fig. 808. *Items which can be easily adapted for use in electronic puzzles and games. These include a box of dowels (upper left), a bell transformer (upper right), a doorbell (lower center) and a toy auto. The auto is operated by a 1.5-volt motor mounted on wheels (shown alongside the auto).*

the main lever, in lowering from the unit, will close the switch during the lowering time as well as during the return movement. Thus, the light or indicator would be tripped during the time the button is depressed. This undesired or premature lighting of the indicator system is prevented by incorporating an additional switch below the indicator button as shown in Fig. 807. This switch (No. 1) is wired in series with the Auto-release switch. The switch beneath the pushbutton is closed only when the pushbutton is not depressed. Thus, when the player pushes the button, the switch beneath the latter is opened and, being in series with the release switch, will prevent the latter from giving an indication In the meantime, the main lever has dropped from the release housing and has moved beyond the switch, closing it. But, as previously mentioned, the first switch under the pushbutton prevents the second switch from functioning. When the player releases the pushbutton, it closes the first switch and therefore permits the second switch to operate. As the main lever returns to its original position it will close the second switch on its return and thus provide the necessary delay.

We have described only one of the methods which can be employed to incorporate a delay system into games and puzzles. The

Fig. 809. *Typical accessories which can be used for making puzzles and games. These are readily available in auto and radio supply stores.*

reader can use this delay technique or let it serve as a basis for devising other delay systems for any of the games or puzzles which he builds. Once the basic puzzles or games have been constructed, they can be improved by the addition of accessories. Also, as the games are played or the puzzles utilized, the con-

Fig. 810. *A motor blinker can be used to add considerable novelty to a game. The motor is a small 1.5-volt unit and operates from a single battery. The drawing at the bottom of page 117 shows a simple speed reduction method.* ➤

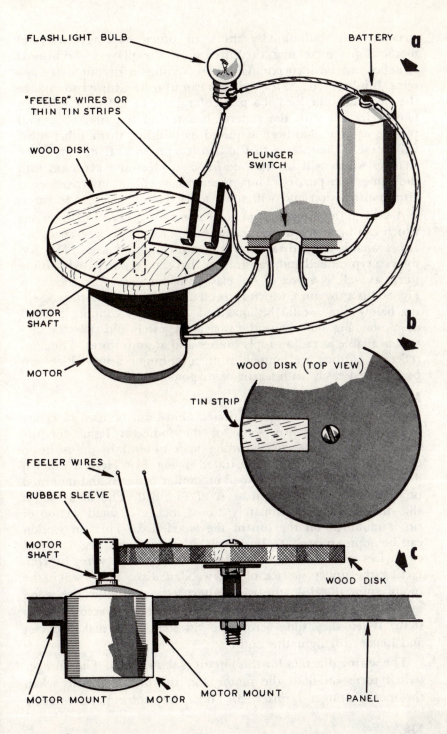

FLASHLIGHT BULB

BATTERY

a

"FEELER" WIRES OR
THIN TIN STRIPS

WOOD DISK

PLUNGER
SWITCH

MOTOR
SHAFT

MOTOR

b

WOOD DISK (TOP VIEW)

TIN STRIP

FEELER WIRES

RUBBER SLEEVE

MOTOR
SHAFT

c

WOOD DISK

MOTOR MOUNT

MOTOR

MOTOR MOUNT

PANEL

structor will undoubtedly think of other improvements and methods for increasing the interest of the players. Additional switches can be incorporated for switching in or out delay systems, blinkers, etc. Extra switches can also be utilized to change the electrical functions of a particular game to add variety. When familiarization with the general layout and function of several puzzles or games has been acquired by building them, other ideas will present themselves to the constructor so that he can make changes which will incorporate his own ideas and processes into the games and puzzles. Thus, the devices will be improved even after construction and will show more personal design features.

A visit to the toy department of a store will reveal many items which can be adapted to puzzles and games. Some accessories are also available at hardware stores or auto supply houses. Fig. 808 shows a typical bell and transformer that can be purchased at dime stores as well as a box of dowels of various sizes. Also shown in Fig. 808 is a toy auto, which runs on a small 1.5-volt battery. Selling below what would be imagined, it actually contains a small dc motor. Fig. 809 shows the variety of jewels and reflector buttons available at radio supply houses and at auto stores. The auto reflector buttons are housed in an aluminum casing which can be removed easily with a pair of diagonal cutters.

Motor blinker unit

A small motor such as mentioned above can be used to create blinking lights. Hence, some of the indicator lights for the puzzles previously discussed can be made to blink by the addition of the motor blinker unit illustrated in Fig. 810. Here, a wooden disk is cut from $\frac{1}{4}$-inch plywood or similar material and mounted on a motor shaft as shown at A of Fig. 810. The diameter of this disk can be approximately 2 or 3 inches. A small section of tin is mounted on the top of the wood disk. This tin section can be approximately $\frac{1}{2}$ by 1 inch. Above the disk are mounted two wires or two strips of tin which act as feeler wires by contacting the upper surface of the wooden disk. Thus, when the motor spins the disk, the wires alternately slide over the wood section of the disk and once every revolution slide across the tin strip. When they slide across the tin strip, they make contact and hence will light the bulb.

The wiring diagram for this circuit is shown at A. The plunger switch turns on both the motor and the blinking bulb when the metal plunger is inserted. Thus, the motor starts rotating and the bulb begins to blink. When the plunger is withdrawn, it removes the voltage from the motor, stopping it. In addition, the voltage is also removed from the bulb so that the latter will not remain lit when the tin strip on the wood disk rests under the feeler wires.

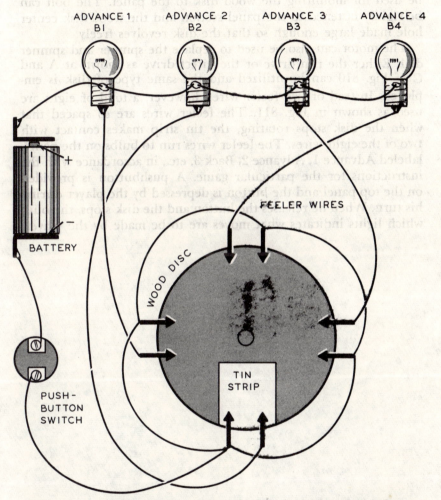

Fig. 811. *The motor blinker unit can be made to operate four lights as shown in this illustration. A pair of feeler wires (thin tin strips) is required for each bulb. Either a pushbutton or metal plunger switch can be used.*

The number of times per minute that the bulb blinks will depend on the diameter of the wood disk, how narrow the tin strip is and how fast the motor turns. If the blinking is too

119

rapid, an indirect drive such as shown at C can be utilized. Here, the motor shaft engages the edge of the disk and hence turns the disk more slowly. The wooden disk must be mounted on the top panel so that it revolves freely. The wiring is the same as shown at A for the motor, battery and bulb. A nut and bolt can be used for mounting the wood disk to the panel. The bolt can be rigidly fastened to the panel section and the wood-disk center hole made large enough so that the disk revolves freely.

The motor can also be used to replace the spinner and spinner dial. Either the rim drive or the direct drive as shown at A and C of Fig. 810 can be utilized and the same type of disk is employed. Instead of two feeler wires, however, a total of eight are used as shown in Fig. 811. The feeler wires are so spaced that when the disk stops rotating, the tin strip makes contact with two of the eight wires. The feeler wires run to bulbs on the panel, labeled Advance 1, Advance 2, Back 3, etc., in accordance with the instructions for the particular game. A pushbutton is provided on the top panel and the button is depressed by the player during his turn. When he releases the button and the disk stops, the bulb which lights indicates what moves are to be made by the player.

construction hints

THE games and puzzles described in this book are quite simple and require the use of a minimum amount of materials. However, even an elementary game, depending upon the flow of an electric current, will not operate if construction is sloppy. The precautions described here should enable the constructor to avoid the necessity for troubleshooting.

Soldering

The primary consideration in soldering is cleanliness. It is very easy to solder to tin but it must be free of grease, oil or paint. Sometimes a scrap piece of tin will appear discolored, a condition that might prevent good soldering. Before spending the time in constructing switches from such a piece of tin, try soldering a small sample to see if it will present any difficulties. If the scrap does not solder well, it would be advisable not to use it. However, even tin that will not otherwise take soldering can be made to do so by scraping it with a file or rubbing it with steel wool.

Soldering iron

A soldering iron rated at 100 watts will do very well, although irons having higher or lower wattage ratings could also be used. A hot soldering iron should not be left where small children could possibly reach it. It is advisable to use a soldering-iron stand, particularly the type that is completely enclosed. This will protect against accidental burns and at the same time minimize the fire hazard caused by having a hot iron rest close to or on a wooden workbench. When the iron is not in use, disconnect it from the power line. To keep the iron in good condition, remove the tip every now and then and shake out the rust that may have accumulated in that part of the iron that holds the tip. If the tip is a screw-in type, its occasional removal will minimize rusting of the threads.

Solder

Solder is made of a mixture of tin and lead in various ratios. Solder with 40% tin and 60% lead is perfectly suitable for working with any of the games and puzzles described earlier. The solder is filled with plastic rosin which acts as a flux. Acid-core solder or solder requiring an external flux can also be used, but rosin-core solder is generally preferred since it is clean and noncorrosive.

Rules for well-soldered connections

1. Make sure the soldering iron is well-tinned. To tin, file or sandpaper the soldering-iron tip. Heat the iron and, when it becomes hot enough, melt some solder on the tip and then rub the tip on a cloth to spread the solder. It may be necessary to repeat the operation a number of times before the tip is completely tinned.

2. Tin both objects to be soldered. If a wire is to be soldered to a metal strip, sandpaper or scrape the wire with a knife if the wire is not already tinned. When the wire is clean, put the tip of the hot soldering iron against it. Apply the solder to the wire and not to the iron. The wire must become hot enough to melt the solder. After tinning the wire, repeat the operation with the object to which the wire is to be soldered. If this object is a piece of scrap tin, you will find that solder will flow very readily on it. Heat the tin with the iron until the tin becomes hot enough to melt the solder.

3. After soldering the two wires or sections together, do not disturb them until the solder has hardened thoroughly. After the joined sections have cooled, give each a slight pull to make sure the soldered joint is secure.

4. Use a minimum amount of solder. Only a small quantity of solder is needed to make a good connection. Large solder blobs can produce trouble since they will often conceal the fact that a joint isn't soldered at all. Also, excess solder can result in short circuits, rapidly discharging the battery or batteries and making them useless.

5. Keep checking the tinned condition of the soldering-iron tip. Retin the iron periodically. When using the iron, wipe

the tip clean with a cloth to keep it in good working condition.

6. Never cool a soldering iron by putting it into water. If you wish to cool an iron rapidly, put the metal shank of the iron between the jaws of a vise or rest the iron on a large metal surface.

7. Solder will splatter once in a while. Keep the iron away from your face. If you get solder droppings into the puzzle or game, remove them, since they may interfere with the operation of one of the switches or could cause a short circuit.

Flashlight bulbs

A variety of flashlight bulbs lend themselves very nicely for use in puzzles and games. If you will look inside the bulb, you will see a tiny wire—the filament. This is the portion of the bulb that lights. The filament is mounted on a small bit of insulating material known as the bead. The color of the bead supplies some information about the characteristics of the bulb.

Some bulbs are shown in Fig. 901 but these are just a few of the many types that could be used. You will find differences

Fig. 901. These bulbs are just a few of the many different types that could be used in the construction of puzzles and games. Solder one wire to the side of the base and another wire to the very tip of the bottom. This will supply the two connecting leads for the bulb.

A   B   C

in the shape of the bulbs and in the bases used (bayonet or screw) while some will come with a built-in lens. Fortunately, these factors are of no consequence in constructing puzzles and games, since connections are made directly to the bulb bases, no sockets being required. However, the voltage and current requirements of the bulbs are important.

The more current used by a bulb (current is measured in amperes), the greater will be the drain on the battery. Hence, in those instances in which several bulbs will be lit at the same time, the load on the battery will be that much greater. If you note, as successive bulbs go on, that the increased load on the battery produces just a dim light, you have an indication that the battery is weak or is not designed to light so many bulbs simultaneously.

The following listing is a brief one and supplies some indication of the bulbs that could be used.

Type No.	Style	Bead Color	Voltage	Amperes
112	B	Pink	1.1	0.22
123	A	Pink	1.2	0.25
136	A	Pink	1.3	0.60
222	B	White	2.2	0.25
PR-8	C	White	1.9	0.60
PR-4	C	Purple	2.3	0.27
PR-6	C	Purple	2.5	0.30

Types such as Nos. 112, 123, etc. are designed for single cells of 1.5 volts. The lower bulb-voltage ratings give a brighter light. When two cells are hooked in series to provide 3 volts, bulbs such as Nos. 222 and PR-4 may be used. Again, the lower voltage rating of the bulbs will provide greater light output.

Dry-cell specifications

The larger cells (batteries) will give longer life, particularly if used with low-current flashlight bulbs such as Nos. 112 and 123. Use smaller cell sizes only where space is at a premium.

When cells are connected in parallel (both plus terminals connected together and both negative terminals connected together), the voltage will still be that of a single cell (1.5 volts) but battery life is doubled.

When cells are connected in series (the plus terminal of one battery connected to the negative terminal of the other), double the voltage is available for use with such bulbs as Nos. 222, PR-4, etc. Use the following listing as a guide in allowing space for the dry cell in the puzzle or game to be constructed. All dimensions shown are in inches.

Type	Length	Diameter
D	2¼	1¼
C	1⅞	1
AA	1¹⁵⁄₁₆	½
K	1¾	⅜

These are just a few of those that are available. Generally, any cell used in a standard-size flashlight is suitable for puzzles and games. Batteries are also identified by particular trade names such as Penlite, Tinylite, Flashlite, etc.

A

AC Buzzers 110
Accessories:
    Game 105
    Puzzle 105
Adding Machine Puzzle 80
Advanced Puzzles 27
Alligator Clips 61
Amount of Solder to Use 122
Auto, Toy 115
Automatic Cable-Release 114

B

Basket-Catch Game:
    Chassis for 100
    Playing Instructions for 100
Batteries:
    in Series 79
    Mounting the 19
    Polarity of 7
    Specifications of 124
Battery Clips 19
Bell Transformer 115
Blinker Unit, Motor 118
Bottom View of the River-Crossing
    Puzzle 33
Bulbs, Flashlight 123
Buzzers:
    AC 110
    Complete Assembly of 107
    DC 105
    Making a 106
    Operating Two from a Single Battery 109

C

Cable-Release, Automatic 114
Cable-Release Units, Camera 113
Camera Cable-Release Units 113
Cell 19
Cells:
    Dimensions of 124
    in Parallel 124
    in Series 124
Chassis for Basket-Catch Game 100
Cleanliness in Soldering 122
Clips:
    Alligator 61
    Battery 19
Coil for Buzzer, Winding a 106
Complex Games 63
Connections between Switches 7
Construction:
    Hints 121
    of the Knife-Switch Puzzle 17
    of Trip-to-the-Shore Game 52
Cross-Country Race:
    Playing Instructions for 68
    Top View of 65
    Wiring Diagram of 66

D

DC Buzzers 105
Delay Systems 112
Diagram of Tri-target Game 97
Dial for Satellite Game, Spinner 93
Dials, Low- and High-Speed Spinner .. 67
Dimensions of Cells 124
Doorbell 110, 115
Double-Indicator Puzzle 21
Dowels 14, 115
Dry Cell Specifications 124

E

Excess Solder 122

F

Flashlight Bulbs 123
Flashlite 124
Front View of Hidden Word Puzzle 88

G

Game:
    Accessories 105
    Basket-Catch 98
    Chassis for Basket-Catch 100
    for Four Players 63
    Hide-and-Seek 47
    Horse Race 101
    Jet Plane 56
    Layout for the Satellite 92
    Layout of Horse Race 101
    Layout of the Tri-target 95
    Playing Instructions for Basket-
        Catch 100
    Playing Instructions for Horse Race 104
    Playing Instructions for Space-Travel 72
    Playing Instructions for Tri-target.. 98
    Puzzle 23
    Puzzle, Wiring Diagram of the 24
    Satellite 91
    Space Travel 69
    Spinner Dial for Satellite 93
    Trip-to-the-Shore 50
    Trip-to-the-Shore, Playing Instruc-
        tions for 56
    Tri-target 96
    Wiring Diagram of Satellite 94
    Wiring Diagram for Space-Travel ... 69
    Wiring Diagram of Tri-target 97
    with Metal Rod Plungers 23
Games:
    Complex 63
    for Several Players 47, 63
    Miscellaneous 91

index

H

Hidden Word Puzzle:
Front View of 88
Wiring Diagram of 89
Hide-and-Seek Game:
Layout of 48
Wiring Diagram of 48
High-Speed Spinner Dial 67
Hints, Construction 121
Horse Race Game:
Layout of 101
Playing Instructions for 104

I

Iron, Soldering 121

J

Jet Plane Game:
Playing Instructions for 60
Wiring for 60
Joining Wires 8

K

Knife Switch:
Puzzle 18
Puzzle, Top View of 18
Puzzle, Wiring Diagram of the 19
Single Pole, Double-Throw 9
Single-Pole, Single-Throw 9

L

Launching a Satellite 91
Layout:
of Adding Machine Puzzle 80
of Hide-and-Seek Game 48
of Horse Race Game 101
of Magic Number Puzzle 85
of Mystery Safe Puzzle 74
of the Satellite Game 92
of the Tri-target Game 95
Little Thinker, Top View of 36
Little Thinker, Wiring Diagram of the.... 37
Low- and High-Speed Spinner Dials 67

M

Machine Puzzle, Adding 80
Magic Number Puzzle 83
Magic Number Puzzle, Layout of 85
Magic Number Puzzle, Wiring Diagram
of the 86
Magnetic:
Switch 83
Switch Puzzle, Variation of 87
Switches, Parallel 103
Switches, Series 103
Switches, Using 102
Magnets in Puzzles, Using 83
Making the Metal-Plunger Switch 12

Metal-Plunger Switch 11
Metal-Plunger Switch, Making the 12
Metal Rod Plungers, Game with 23
Metal Tubing, Using 13
Miscellaneous Games 91
Miscellaneous Puzzles 73
Motor, 1.5-Volt 115
Motor Blinker Unit 118
Mounting the Battery 19
Mystery Safe Puzzle 73
Mystery Safe Puzzle, Layout of 74
Mystery Safe Puzzle, Wiring Diagram of
the 75

N

Number Puzzle, Magic 83

P

Parallel Cells 124
Parallel Magnetic Switches 103
Penlite 124
Players:
Game for Four 63
Games for Several 47, 63
Playing Instructions:
Basket-Catch Game 100
Cross-Country Race 68
Horse Race Game 104
Jet Plane Game 60
Space-Travel Game 72
Trip-to-the-Shore Game 56
Tri-target Game 98
Plunger Switches:
Making the 12
Top Mounts for 15
Working with 45
Plungers Game with Metal Rod 23
Polarity of Batteries 7
Prospecting Puzzle 86
Pushbutton:
Switch 9, 10
Type of Puzzle 21
Puzzle:
Accessories 105
Adding Machine 80
Bottom View of the River-Crossing.... 33
Double-Indicator 21
Front View of Hidden Word 88
Game 23
Hidden Word 87
Knife-Switch 17
Layout of Adding Machine 81
Layout of Magic Number 85
Layout of Mystery Safe 74
Magic Number 83
Mystery Safe 73
Prospecting 86
Pushbutton Type of 21
River-Crossing 30
Solving the River-Crossing 35
Street-Light 28
Switch Construction Details for the
River-Crossing 32
Top View of the River-Crossing 31
Puzzle Wiring Diagram:
Magic Number 86
Mystery Safe 75
River-Crossing 34
Street-Light 28
Puzzles:
Advanced 27
Miscellaneous 73
Simple 17
Using Thumbtacks in 74

R

Race:
Cross-Country 63
Game, Horse 101
Playing Instructions for Cross-
Country 68
Top View of Cross-Country 65
Wiring Diagram of Cross-Country 66
Relays 111
Release:
Automatic Cable- 114
Units, Camera Cable 113
River-Crossing Puzzle:
Bottom View of the 33
Solving the 35
Switch Construction Details for the 32
Top View of 31
Wiring Diagram of the 34
Rotary Switch 11
Rotary Tin Strip 74
Rules for Well-Soldered Connections 122

S

Safe Puzzle, Mystery 73
Safety 46
Satellite:
Game 91
Game, Layout for 92
Game, Wiring Diagram of 94
Launching a 91
Series, Batteries in 79
Series, Cells in 124
Series Magnetic Switches 103
Simple Puzzles 17
Single Battery, Operating Two Buzzers
from 109
Single-Pole Single-Throw Knife Switch.. 9
Single-Pole Single-Throw Knife Switch.... 9
Solder 122
Solder, Amount to Use 122
Solder, Excess 122
Soldered Connections, Rules for 122
Soldering:
Cleanliness in 122
Iron 121
Iron, Tinning the 122
Wire to Switches 8
Wires 122
Solving the River-Crossing Puzzle 35
Space-Travel Game 69
Space-Travel Game, Playing Instructions
for 72
Space-Travel Game, Wiring Diagram for 69
Spinner Dials:
for Satellite Game 93
Low- and High-Speed 67
Marking the 55
Trip-to-the-Shore Game 55
Using Two 65
Street-Light Puzzle 28
Street-Light Puzle, Wiring Diagram of
the 28
Substituting Switches 20
Switches:
Connections Between 7
Construction Details for the River-
Crossing Puzzle 32
Double-Pole, Double-Throw 9
Knife 10
Magnetic 83
Metal-Plunger 11
Pushbutton 9, 10
Rotary 11

Single-Pole, Single-Throw 9
Soldering Wires to 8
Substituting 20
Toggle 9, 11
Tri-target 96
Wafer 9
Wood-Plunger 14
Systems, Delay 112

T

Thumbtacks, Using in Puzzles 74
Tinning the Soldering Iron 122
Tinylite 124
Toggle Switch 9, 11
Top Mount for Plunger Switches 15
Top View:
Cross-Country Race 65
Little Thinker 36
River-Crossing Puzzle 31
Toy Auto 115
Transformer and Doorbell 110
Transformer, Bell 115
Trip-to-the-Shore Game:
Construction of 52
Playing Instructions 56
Spinner Dial for 55
Top View of 49
Wiring Diagram of 53
Tri-target Game:
Layout of 95
Playing Instructions 98
Switches for 96
Wiring Diagram of 97
Twenty-one, Wiring Diagram for 43

V

Voltage Divider 79

W

Wafer Switch 9
Wires:
Joining 8
Size 8
Soldering 122
Wiring Diagram:
Cross-Country Race 66
Double-Indicator Puzzle 22
Game Puzzle 24
Hidden Word Puzzle 89
Hide-and-Seek Game 48
Jet Plane Game 60
Knife-Switch Puzzle 19
Little Thinker 37
Magic Number Puzzle 86
Mystery Safe Puzzle 75
River-Crossing Game 34
Satellite Game 94
Space-Travel Game 69
Street-Light Puzzle 28
Tri-target Game 97
Trip-to-the-Shore Game 53
Twenty-one 43
Wood-Plunger Switch 14
Wood-Plunger Switch, Constructing the 15
Word Puzzle, Hidden 87
Working with Plunger Switches 45

PRINTED IN THE UNITED STATES OF AMERICA